Microbial Control of Plant Pests and Diseases

Aspects of Microbiology

Series editors
Dr. J.A. Cole, University of Birmingham
Dr. C.J. Knowles, University of Kent
Dr. D. Schlessinger, Washington University School of Medicine, USA

This series brings important topics of current interest, conceptual difficulty or controversy to the attention of second and third year undergraduates, and to more senior scientists wishing to acquaint themselves with recent developments in fields outside their own specialities. One of the principal objectives of the series is to bridge the gap between introductory texts and research literature. The series editors and the publisher have sought to achieve this objective by providing brief, inexpensive texts which will familiarize students with current research using language and illustrations carefully controlled to conform to undergraduate expectations.

1. Oral Microbiology *P. Marsh*
2. Bacterial Toxins *J. Stephen and R.A. Pietrowski*
3. The Microbial Cell Cycle *C. Edwards*
4. Bacterial Plasmids *K. Hardy*
5. Bacterial Respiration and Photosynthesis *C.W. Jones*
6. Bacterial Cell Structure *H.J. Rogers*
7. Microbial Control of Plant Pests and Diseases *J. Deacon*
8. Methylotrophy and Methanogenesis *P.J. Large*

Aspects of Microbiology 7

Microbial Control of Plant Pests and Diseases

Dr. J.W. Deacon
Lecturer in Microbiology
University of Edinburgh

Van Nostrand Reinhold (UK) Co. Ltd.

TO DENIS GARRETT
in appreciation of his friendship and guidance and in recognition of his contribution to microbial control of plant disease

First published 1983

ISBN: 0 442 30512 5

Published by Van Nostrand Reinhold (UK) Co. Ltd.
Molly Millars Lane, Wokingham, Berkshire, England

Printed and bound in Hong Kong

Contents

Contents

Preface

The use of micro-organisms to control crop pests and diseases is an exciting and rapidly advancing branch of applied biology. The topic has previously been considered separately in terms of pests and diseases, but this is the first book that gives equal weight to the two themes and makes comparisons between them. My aim throughout has been to show that there is an intimate relationship between 'pure' and 'applied' aspects of the subject; so the reader will see several cases in which the development of a practical control measure must await further understanding of some basic phenomenon, as well as examples in which study of an existing control measure has revealed information of fundamental significance.

The book is intended to appeal to microbiologists, plant pathologists and entomologists, but also to anybody with an interest in the applications of biology. With this in mind I have been careful to explain any technical terms as they arise and also to expand on any concepts that might not be familiar to the general biologist. My approach has been to consider in depth some examples of microbial control that are in current commercial practice and to use them to derive and illustrate general principles. In each case I have given some background to the pest or disease problem and discussed why microbial control is appropriate; then the control measure is described, together with the mode of action at the fundamental level, and the outstanding problems and future prospects are discussed. Inevitably in a book of this size I have had to omit some topics, of which the potential use of protozoa as control agents is perhaps the most obvious instance; but the themes and general principles can be extended to cover any excluded topics.

I am grateful to all those people who allowed me to use published material, and particularly to Dr Terry Brokenshire, Mr Howard Chittick (Fairfax Biological Laboratory Inc.), Dr Joan Webber and Dr Neil Wilding who supplied material especially for the book. It is a pleasure to thank Dr Peter Jeffries and Professor Denis Garrett who read the typescript for me, and the series editors who made many useful suggestions. However, I alone am responsible for any errors or omissions.

March 1982 *Jim Deacon*

1 Introduction

Society is faced with a dilemma. On the one hand there is an ever-increasing need for food and especially for improved crop production in the developing countries. On the other hand some of the methods currently used to achieve higher yields, especially by pest and disease control, are environmentally undesirable. In the search for safer and more lasting methods, biologists have turned their attention to the possibility of using other organisms as *biological control agents*, and microbiologists are contributing in the development of *microbial control agents*. The aim of this book is to discuss some of the successes of microbial control and some of the main prospects for future development, and also to show how these are inextricably linked with research into the fundamental or 'pure' aspects of microbiology. For reasons of space I have excluded discussion of the microbial control of weeds, though weeds are important factors limiting crop yields; good recent reviews of this topic are given by Templeton *et al.* (1979) and Hasan (1980). Also I have purposefully fully excluded control of diseases of man and other vertebrates by vaccination (though this might validly be termed microbial control) because it is most appropriately discussed in medical and veterinary texts.

Microbial control and biological control: definitions and scope

There is much disagreement on what constitutes biological control. DeBach (1964) defined it as '*the action of predators, parasites, or pathogens in maintaining another organism's population density at a lower average than would occur in their absence*'. This was meant specifically in relation to insect pests and weeds: as we shall see it is not broad enough to cover plant pathogens, control of which is often brought about by competition, among other mechanisms. On the other hand some plant pathologists have adopted an extremely broad view of biological control and include, for example, all cases in which a plant is bred for disease-resistance, i.e. by genetic manipulation. Baker & Cook (1974), for instance, define biological control as '*the reduction of inoculum density or disease-producing activities of a pathogen or parasite in its active or dormant state, by one or more organisms, accomplished naturally or through manipulation of the environment, host, or antagonist, or by mass introduction of one or more antagonists*'. It cannot be denied that any workable control measure is a valuable one, but the best definition must be one that circumscribes an approach to control, and Baker & Cook's definition does little more than exclude some cases of chemical control.

The definition that I propose is based on one by Garrett (1970) but covers both pests and pathogens of plants. *Biological control is the practice in which, or process whereby, the undesirable effects of an organism are reduced through the agency of another organism that is not the host plant, the pest or pathogen, or man.* In other words biological control is mediated by a 'third party'; in the case of microbial control this is a micro-organism.

It is useful to dwell on this theme for a while. A simple and indisputable case of biological (microbial) control would involve adding one organism (micro-organism) to control another. There are many such cases in current practice and a few are mentioned below. The bacterium *Bacillus popilliae* is added to grass turf in the USA to control larvae of the Japanese beetle (chapter 2); a virus has been sprayed on to spruce trees in Canada to control larvae of the European spruce sawfly (chapter 3); spores of the fungus *Beauveria bassiana* are sprayed on to potato crops in the Soviet Union to control the Colorado beetle (chapter 4). Turning to plant pathology, the antagonistic fungus *Peniophora gigantea* is added to pine stumps in Britain to prevent them from being colonized by the aggressively pathogenic fungus *Heterobasidion annosum* (chapter 5); attenuated strains of some viruses (e.g. tobacco mosaic virus) are used to protect crops against virulent strains (chapter 5); the bacterium *Agrobacterium radiobacter* is used in a root-dip treatment to protect many woody plants against the crown gall bacterium *Agrobacterium tumefaciens* (chapter 5).

Not all examples of biological control are so straightforward. If cereals are grown continuously (i.e. year after year) on the same site they may at first be heavily diseased by the take-all fungus but the level of disease subsequently declines. Cereals can be grown profitably on these 'take-all decline' sites, and this is done in commercial practice. We know that micro-organisms are responsible for take-all decline and we are exploiting them, so this is a valid example of microbial control. Ironically, however, we still are not sure which micro-organisms are responsible (chapter 7).

A further example: in the glasshouse cropping industry it is common practice to partly sterilize soils by steam-air mixtures or chemical fumigants but the soils are intentionally not completely sterilized (chapter 6). Partial sterilization is sufficient to kill most weed seeds and plant pathogens but it leaves some of the resident saprophytic micro-organisms, which include fungi and bacteria that antagonize plant pathogens. The antagonists prevent the soil from becoming recolonized by any pathogens that persist in local pockets. This is an example of microbial control even though it may involve the use of chemicals. But strictly speaking it is included in the group of practices termed *integrated control*, i.e. practices in which more than one control measure is operating. In this case there is a primary and direct effect of heat or chemicals on the plant pathogens and a secondary, indirect effect operating through the activities of micro-organisms.

Even plant breeding can be included in our definition of biological control if (and only if) the new plant cultivar is resistant to disease because it supports a population of controlling organisms. There is no clear-cut example of this at present though there are some likely candidates. For example, in Canada spring wheat can be made resistant or susceptible to the disease termed common root rot by substitution of the chromosome pair 5B from a resistant or susceptible parent cultivar. This is correlated with changes in the root microflora though there may not be a causal relationship (chapter 6).

Some types of control must be excluded from our definition even though they involve micro-organisms. Returning to the example of attenuated strains of viruses to protect against virulent ones, this would be excluded from the definition if it could be shown that the virus has some component (say, a surface component) that is recognized by the plant and induces resistance and if this component alone could be applied to give the same effect. A similar case would be if a product of a micro-organism (like an antibiotic) is alone responsible for control and if this product can be manufactured and used. An example approaching this is seen in the use of *Bacillus thuringiensis* to control larvae of the Lepidoptera (chapter 2). The bacterium is grown in culture and produces a mixture of spores and toxin crystals which together constitute the control agent. In many cases the crystals alone can be used to give control, but the bacterium is recognized to play a secondary role in the control process by invading hosts previously weakened by the toxin, so this is covered by our definition. One of the reasons sometimes given for including plant breeding as a method of biological control is that in the foreseeable future we may be able to incorporate microbial genes into the plant genome to achieve control. This is true particularly in the case of crown gall disease (chapter 5). However it must be excluded from our definition because it is purely genetic manipulation, irrespective of where the genes come from.

In summary of this section, microbial control can be achieved in several different ways but all of the cases that we include in the definition have one thing in common: *an integral part of the control process is the activity of a micro-organism*. So, in effect, microbial control is one of the practical applications of microbial ecology.

Background: the need for pest and disease control

Table 1 gives estimates of the annual crop losses on a global scale caused by insect pests, plant diseases and weeds. Such figures are difficult to obtain in practice (Cramer, 1967) and may not be accurate, but similar figures are given in various sources so they are probably of the right magnitude. First it should be noted that the figures refer to losses incurred after existing control measures have been applied (though of course they are not always applied and certainly not always applied most effectively). Second, the problems are not confined to the poorer countries: although losses tend to be greater in the Third World (see, for example,

Table 1 Estimated annual crop losses caused by insect pests, plant diseases and weeds (From Cramer (1967) by kind permission of the author and publisher)

Commodity	Value ($1000 million)*		Losses (%) due to			Total loss
	Actual	Potential	Insects	Diseases	Weeds	(%)
Wheat	18.5	24.3	5.0	9.1	9.8	23.9
Rice	19.6	36.4	26.7	8.9	10.8	46.4
Maize	11.4	17.5	12.4	9.4	13.0	34.8
Other cereals	11.4	19.8	6.6	8.6	12.1	27.3
Potatoes	10.6	15.6	6.5	21.8	4.0	32.3
Sugar	7.6	13.9	16.5	16.5	12.2	45.3
Vegetables	16.7	23.1	8.7	10.1	8.9	27.7
Fruit	14.3	20.1	5.8	16.4	5.8	28.8
Stimulants	7.2	11.4	11.4	14.9	10.5	36.8
Oil crops	10.6	15.7	11.5	10.2	10.8	32.5
Fibres/rubber	8.6	12.7	14.2	11.8	6.3	32.3

* Year 1965

figures for sugar and rice) estimated losses of about 20 to 30% occur in potatoes and wheat which are grown mainly in the agriculturally advanced communities. A third point of relevance tends to be obscured in the summary data: some individual pests and pathogens cause major crop losses (see Table 4, p. 44) but most of the figures in Table 1 result from the activities of a range of pests and pathogens. Consequently there is no single dramatic cure; each pest or disease problem must be considered separately and any control measure must be related to local circumstances, such as climate, traditional practices, availability of labour or finance, etc. Some of the examples in the text are therefore of restricted relevance, but they can be used to illustrate and arrive at general principles.

Microbial control in relation to other control methods

Some of the main advantages of microbial control methods can be listed as follows.

1 They are 'safe' in comparison with some of the more generally toxic chemical control agents, and in particular the microbes are not accumulated in food chains. However, the safety testing of some insect pathogens is very expensive, and safety considerations have until recently seriously limited the application of some control agents – viruses in particular (chapter 3). It is worth stating the obvious: that living organisms can replicate, unlike chemicals, and they might prove impossible to contain if they are subsequently found to be hazardous. The myxoma virus of rabbits is a clear example of a non-containable control agent (chapter 3) though in this case there is no evidence of any potential hazard.

2 Microbial control agents can be persistent, giving lasting control. This does not apply in all cases but certainly there are examples in which a microbial control agent, once introduced, has contained a pest problem at a sub-economic level (chapters 2 and 3). It is often argued that biological control methods are more permanent than are chemical control methods or the use of disease-resistant plant cultivars because they are broadly based (so more than a single mutation is needed to overcome them) and because they are often aimed at a fundamental aspect of the pest's or pathogen's biology that is not readily alterable. In some cases these arguments are justified and it is equally true that pests or pathogens have overcome some of the more widely used chemicals or disease-resistant plants. However, direct comparison is unreasonable because no microbial control agent has yet been used on the scale of some other control methods. In fact, there are some examples of the breakdown of microbial control methods (chapters 2 and 3).

3 Microbial control agents, and biological control agents in general, have at most slight effects on the ecological balance. In particular, they do not eradicate the 'natural enemy complex', i.e. all those organisms that naturally help to keep a pest or pathogen in check. This is an important point because many chemical control agents affect 'non-target' as well as 'target' organisms and there are several examples in which control of one pest or disease by a chemical has resulted in the emergence of another one as a result of the destruction of natural enemies (chapter 6). This is termed pest or disease 'trading'. Alternatively (and for the same reason) a 'boomerang effect' can occur, in which the pest or pathogen re-emerges at an even higher level than before once the effect of a chemical has worn off. Entomologists have recently shown great interest in the field of 'Insect Pest Management' – an approach that makes use of natural enemies and supplements their effects, if necessary, with other control measures; the aim is to achieve a stable low level of pest damage, and microbial control agents are particularly useful in this respect. Plant pathologists in general have been slower to embrace this idea, mainly because the necessary background work has yet to be done.

4 Microbial control agents are often compatible with other control agents, including chemicals, and can be used in conjunction with them. This is probably the most realistic approach to microbial control: it is unrealistic to think that microbial control agents will ever completely replace the existing control methods.

There are a number of disadvantages of microbial control; they are mentioned in relation to specific examples in the following chapters and summarized in chapter 8. The general disadvantages include the long-term study that is necessary before a microbial control method can be perfected, especially in relation to the scale on which it is subsequently going to be used, and the fact that micro-organisms cannot easily be patented; industry is therefore reluctant to invest in the development programmes. Indeed, the ideal of biological control – that it should be tailored specifically to the pest or disease problem in hand – is often incompatible with the activities of large companies.

Practical approaches to microbial control

It is useful to distinguish four approaches to microbial control, though they may overlap considerably in practice.

1 Introduction or mass-release This involves introducing a control agent into an area from which it was previously absent. In appropriate conditions it will then proliferate and spread, bringing about a lasting degree of control. This is the cheapest and most effective way of reducing permanently a pest or disease problem and it is practised when the pest or pathogen is known to have been introduced into a new area from an endemic one. The rationale, which experience shows to be correct, is that a pest or pathogen becomes a problem when it is separated from the natural enemy complex that normally contains it in its area of origin, so the balance can be restored by introducing the natural enemies. Successful examples of the approach include the introduction of insect-pathogenic viruses into North America to control forest pests (chapter 3) and the mass release of *Bacillus popilliae* to control the Japanese beetle in the USA (chapter 2).

2 Microbial pesticides This approach involves using a micro-organism exactly as one would a chemical control agent, i.e. repeatedly as and when necessary. It is used for microbes that do not persist in the field or, at least, that do not persist at high enough levels to give satisfactory control once a problem begins to develop. Examples include the use of *Bacillus thuringiensis* (chapter 2), *Peniophora gigantea* (chapter 5) and *Agrobacterium radiobacter* (chapter 5).

3 Exploiting natural microbial control This approach relies on identifying the agents that naturally contain a pest or pathogen and then preserving the conditions that favour control. It is control by minimal interference and therefore it may seem ironical to be included as a practical approach. A good example is the exploitation of decline soils like those exhibiting take-all decline (chapter 7). Much research is currently being done on these to see whether they can be created without the need to go through a disease peak and also to see if 'break crops' can be used occasionally in a monoculture, to introduce more flexibility into a cropping system (and to control weeds etc.) without losing the decline factor.

One practical way of maintaining a naturally occurring microbial control is to avoid the use of chemicals likely to destroy it. This in turn requires that we can identify the important natural microbial control agents and test the effects of chemicals on them. Chapter 4 discusses one such example.

A general problem with naturally occurring microbial control, as in the case of introductions mentioned under (1) above, is that there is often a degree of crop damage or a periodic resurgence of the problem. This is especially true if the control agent depends on the pest or pathogen (e.g. a virus needs a host) and we see a cyclic rise and fall of the pest or pathogen followed by a corresponding rise or fall in the level of its control agent (Figure 1). Another reason for this is that a minimum host density is necessary for the progressive spread of a disease, so an insect pathogen, for example, cannot cause an *epizootic*[1] until the insect itself is more or less uniformly distributed. In practice, a degree of damage is acceptable in some crops (e.g. forest trees) but not in others (e.g. soft fruit and ornamentals).

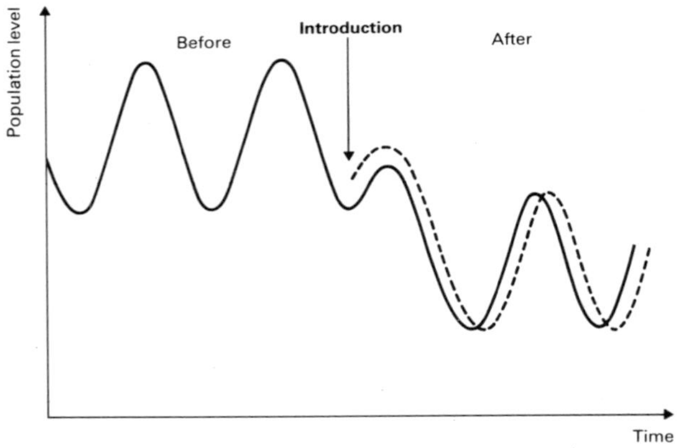

Fig. 1 Population level of a pest (solid line) before and after introduction of a microbial control agent. The broken line indicates the population level of a control agent that *depends* on the pest.

4 Manipulation of the environment To some extent this approach can be coupled with (2) and (3) above but it also merits comment in its own right. Sometimes a change in husbandry practices is all that is needed to effect microbial control, by

[1]Strictly speaking, the term *epidemic* refers to spread of disease in human populations, so disease spread among animals is *epizootic*, and that among plants is *epiphytotic*.

enabling the control agent to proliferate or by predisposing the pest or pathogen to its effects. Attention is currently being given to the effects of pesticides in stressing insects so that they are more susceptible to microbial control agents (chapter 4). Likewise, soil treatment with a sub-lethal dose of a fumigant like carbon disulphide or methyl bromide predisposes the fungus *Armillariella mellea* (a serious pathogen of trees) to attack by antagonistic fungi like *Trichoderma* spp. (chapter 6). Food microbiologists have long recognized the selective effects of different physical and chemical treatments on the spoilage microflora; exactly the same applies to natural and agricultural environments except that, with sufficient knowledge of the vastly more complex community, one hopes to reduce the populations of undesirable organisms whilst enhancing those of potential control agents.

The plan of the book

The rest of this book describes some practical approaches to microbial control. In the case of pests these are dealt with under the headings: use of bacteria, viruses and fungi, because different background information is needed in each case. Control of diseases is dealt with in terms of the approaches used in practice, because the approaches themselves require some background knowledge. Finally the methods of pest and disease control are compared in chapter 8, where the special constraints on microbial control in general are discussed.

References

BAKER, K. F. and COOK, R. J. (1974). *Biological Control of Plant Pathogens*. Freeman, San Francisco.

CRAMER, H. H. (1967). *Plant Protection and World Crop Production*. Farbenfabriken Bayer AG, Leverkusen.

DEBACH, P. (1964). *Biological Control of Insect Pests and Weeds*. Reinhold, New York.

GARRETT, S. D. (1970). *Pathogenic Root-Infecting Fungi*. University Press, Cambridge.

HASAN, S. (1980). Plant pathogens and biological control of weeds. *Review of Plant Pathology* 59: 349–56.

TEMPLETON, G. E., TE BEEST, D. O. and SMITH, R. J. (1979). Biological weed control with mycoherbicides. *Annual Review of Phytopathology* 17: 301–10.

General textbooks

Entomology: RICHARDS, O. W. and DAVIES, R. G. (1977). *Imms' General Textbook of Entomology. 10th Edition*, Chapman & Hall, London.

Plant Pathology: AGRIOS, G. N. (1978), *Plant Pathology. 2nd Edition*, Academic Press, New York.

Nematology: DROPKIN, V. H. (1980). *Introduction to Plant Nematology*, Wiley, New York.

2 Microbial control of pests: use of bacteria

Although 100 or so bacteria cause disease of insects (Cantwell, 1974) only a few are used commercially as control agents. The main ones are *Bacillus thuringiensis* and a group of bacteria classified as varieties of *Bacillus popilliae*. They are amongst the most successful of all microbial control agents and they differ from one another in several interesting respects, not least of which is the fact that *B. thuringiensis* is used as a microbial insecticide whereas *B. popilliae* is used for introductions or mass-release. They are treated separately in detail and their features will then be compared at the end of this chapter.

Bacillus thuringiensis

This bacterium was first recorded in 1901 as the cause of the damaging 'sotto' disease of silkworms in Japan. It was again isolated in 1927 by Mattes in Germany and given the name *B. thuringiensis*. Mattes's strain was tested as a microbial control agent by several workers with encouraging results, and by the mid-1940s the first commercial preparation, named 'Sporiene', was available in France. This gave very variable results in practice and it was not until 1958, when the problems of variability had been overcome, that the first commercial product was released for field use in the USA. Since then *B. thuringiensis* has been marketed under several trade names in many countries for control of a wide range of pests. It is currently perhaps the most widely used and commercially the most profitable microbial control agent.

B. thuringiensis shows several interesting features. The bacterium itself is only weakly pathogenic but at the time of sporulation it produces an intracellular protein crystal which contains a stomach poison. The vegetative cells lyse to release a mixture of spores and crystals (one per cell) which together constitute the microbial insecticide. The mixture is ingested by insect larvae feeding on contaminated vegetation; the toxin causes gut paralysis and the bacterium is then able to invade the weakened host and cause a lethal septicaemia. In fact the initial variability of the French product Sporiene was probably due to a lack of appreciation of the importance of the toxin in the disease process. A second interesting feature is the specificity of *B. thuringiensis*, which affects only larvae of members of the Lepidoptera (butterflies and moths) and even then not all members of that order. Only these insects have a sufficiently high pH in the mid-gut to solubilize the protein crystal and release the toxin; all other insects and vertebrates are unaffected by the bacterium. It has therefore been given clearance for use on food crops because of its complete safety, and it has the further advantage (unusual amongst microbial control agents) of being very fast acting. However, the spores and crystals are inactivated by prolonged exposure to the UV component of sunlight, so they tend not to persist on vegetation, and the control agent has to be applied repeatedly during the season.

The bacterium and its toxin *B. thuringiensis* is a Gram-positive spore-forming rod, 3 to 5 × 1.0 to 1.2 μm, sometimes motile by lateral flagella. It grows well in a range of simple laboratory media like nutrient broth, in aerated conditions and within the temperature range 15 to 40°C (optimum about 30°C). Commercially it can be produced cheaply on substrates like soybeans and molasses. Fifteen serotypes are currently recognized, based on the flagellar or 'H' antigens, and they differ further in biochemical properties and host range. They are given varietal names as follows:

Serotype[1]	Varietal name[2]	Serotype[1]	Varietal name[2]
1	*thuringiensis*	8a,8b	*morrisoni*
2	*finitimus*	8a,8c	*ostriniae*
3a	*alesti*	9	*tolworthi*
3a,3b	*kurstaki*	10	*darmstadiensis*
4a,4b	*sotto, dendrolimus*	11a,11b	*toumanoffi*
4a,4c	*kenyae*	11a,11c	*kyushuensis*
5a,5b	*galleriae*	12	*thompsoni*
5a,5c	*canadensis*	13	*pakistani*
6	*entomocidus,*	14	*israelensis*
	subtoxicus	15	*indiana*
7	*aizawai*		

[1]Subdivided where only some of the antigens are common.
[2]Different names for one serotype reflect different biochemical properties.

In most cultural conditions cells of *B. thuringiensis* sporulate readily as they age and at the same time produce a parasporal body outside the exosporium membrane (Figure 2). This parasporal body is the proteinaceous toxin crystal. Only serotype 2 differs appreciably from the rest in forming the crystal inside the exosporium. The cells then lyse, releasing the spores and crystals into the culture medium, from which they can be separated by centrifugation and made into an insecticide. Most commercial preparations are dusts or wettable powders, though liquid suspensions are becoming more common. All contain a mixture of spores and crystals.

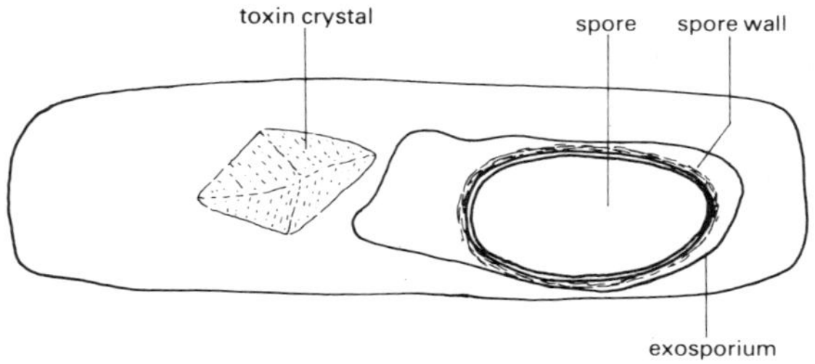

Fig. 2 *Bacillus thuringiensis.*

Because of their importance the crystals have been studied intensively (Fast, 1981). They are normally bipyramidal (though other shapes occur) and they comprise up to 30% of the cell dry weight. They are composed entirely of protein, so they are heat-labile, and they are insoluble in water or organic solvents but readily solubilized in alkaline reducing conditions. By electron microscopy and X-ray diffraction analysis they are seen to have a regular sub-unit structure of rod- or dumb-bell-shaped bodies each about 15 × 5 nm. On digestion with trypsin a range of different polypeptides are produced but always there is a *protease-resistant peptide* of about 60 000 molecular weight which has been shown to be toxic. It is termed δ-*endotoxin* and it is the basis of the specificity of *B. thuringiensis* towards the Lepidoptera. Thus the crystal itself is a protoxin from which the active material is released by proteolysis. The origins of the crystal proteins within the cell are unclear, though the amino acids are known to be made available by recycling cellular protein in the early stages of sporulation. The crystals show antigenic similarities with some of the spore wall components and on this basis they have been suggested to arise by over-production of the spore wall components themselves.

In addition to δ-endotoxin, most strains of *B. thuringiensis* also produce a range of other toxic metabolites, including one true toxin termed β-*exotoxin* or *thuringiensin*. This is an ATP analogue of molecular weight 701 and is a specific inhibitor of DNA-dependent RNA polymerase, competing with ATP for a binding site on the enzyme. It is extremely toxic to a range of insects and also to some vertebrates, and it must be eliminated from commercial preparations of *B. thuringiensis*. This can be done easily by washing because thuringiensin is water-soluble, but in fact most commercially used strains of the bacterium have been selected not to produce it. There has been some interest, especially in the Soviet bloc, in its possible use as a control agent, but it is unlikely ever to get clearance from the appropriate registering authorities in the West.

Mode of action of *B. thuringiensis* The toxin crystals are very stable under most conditions and if they are ingested by most insects or by vertebrates they either pass unchanged through the gut or are inactivated by the acidic conditions in the stomach. Many lepidoptera have a highly alkaline mid-gut (pH 10.2 to 10.5) and in these conditions the crystals dissolve (minimum pH 8.9) and the toxin is released by proteolytic enzymes. There is some evidence that the enzymes of different insects release different polypeptides from the crystals, so further specificity may arise in this way.

Affected species can show two separate responses to the toxin. In all cases there is rapid gut paralysis which occurs within an hour or so of ingestion. The insect then stops feeding and is no longer a problem. A number of degenerative changes are seen in the gut epithelium and these are accompanied by metabolic breakdown such that ions leak from the gut lumen into the haemolymph. A few insects, like the silkworm, show a more pronounced response in that general paralysis occurs soon after the toxin is ingested. Although this is unimportant from the viewpoint of control it must nevertheless be satisfying to the farmer. It is associated with a rapid rise in blood pH, owing to equilibration with the gut contents, and the effect can be reproduced experimentally simply by injecting alkali.

Many insects are killed by the toxin crystal alone, but in other cases a combination of spores and crystals is necessary and this combination is always present in the commercial formulation. The spores are prevented from germinating at pH values above 9.0 but they germinate when the gut pH is lowered by equilibration with the blood. The vegetative cells then invade the tissues and cause a lethal septicaemia.

Applications for microbial control By 1971 commercial preparations of *B. thuringiensis* were registered in the USA for use on more than 20 agricultural crops and on many trees and ornamental plants (Falcon, 1971). Registered applications included control of (1) the alfalfa (lucerne) caterpillar *Colias eurytheme*, (2) the cotton bollworm *Heliothis zea*, (3) the cabbage looper *Trichoplusia ni*, (4) the cabbage white butterfly *Pieris rapae*, (5) the tobacco budworm *Heliothis virescens*, (6) the gypsy moth *Lymantria dispar* and (7) the European corn borer *Ostrinia nubilalis*. Registration of a product in the USA requires proof not only of safety but also of effectiveness when compared with existing control agents, so *B. thuringiensis* has clearly 'made the grade'. Several varieties are now used commercially, the important features in this respect being firstly their somewhat different spectra of activity against members of the Lepidoptera and secondly the selection of strains with least effect on beneficial insects like the honey-bee and silkworm.

Advantages and disadvantages of *B. thuringiensis* The main advantages of *B. thuringiensis* as a control agent can be listed as follows. (1) It has a broad spectrum of activity against larvae of the Lepidoptera (about 300 in all) and this order includes some of the most serious crop pests. A broad spectrum of activity is desirable for the commercial producer of a pesticide because it represents a large potential market, and correspondingly more time and money can be spent on development programmes. (2) *B. thuringiensis* has no effect on vertebrates, including man, and it has a generally negligible effect on insects that form part of the natural enemy complex of pests. (3) The ease of production makes the product competitive in price with conventional chemical pesticides. (4) It gives rapid control just like the chemical pesticides. (5) It is relatively stable during storage, especially as a powder. (6) No resistance has been reported despite the relatively large scale on which it has been used.

Against these must be set the disadvantages. (1) The spores and, to a lesser degree, the crystals are inactivated by prolonged exposure to UV radiation. This can be overcome in part by the use of UV-protectants like carbon but still it limits the life of the product on the plant. (2) The spores and crystals remain on the plant surface and therefore are effective only against surface-feeding insects or the surface-feeding stages of boring and burrowing insects. (3) The spores and crystals show a general lack of persistence in the environment, at least at levels that can achieve lasting control of a pest problem. Ironically this is a welcome feature to the manufacturer, for whom too much success in that respect spells bankruptcy!

One of the last important features for mention is the compatibility of *B. thuringiensis* with other control agents including insecticides. Indeed a classic example of its use is for control of the cabbage looper on cabbage and lettuce crops in Arizona and California. These crops must be protected from the pest right up until the time of harvest and yet the most effective pesticides leave unacceptable residues on the crop and their use must be discontinued one month before harvest. At this stage they are replaced by *B.thuringiensis* which can be applied until the harvest date, if necessary.

Outstanding problems and future research More information is needed on the δ-endotoxin and the way in which it is released from the crystal in the insect gut. This might pave the way for products with narrower spectra of activity and more closely tailored to specific needs in pest control. Coupled with this there is a need for detailed study of the genetics of *B. thuringiensis*, with a view to strain construction in the future. Of particular interest is the mounting evidence that crystal production might be coded by plasmid genes. For example Ermakova *et al.* (1978) found that three plasmids were associated with crystal production: they disappeared from cells grown in media that supported sporulation but not crystal-formation and they reappeared when the cells were transferred to media that supported both crystal and spore formation. The plasmids were suggested to arise by specific excision and amplification of chromosomal genes. Such studies raise the possibility that genes coding for crystal production might be transferred from *B. thuringiensis* into other bacteria, including saprophytic ones that would be able to grow in the absence of insect hosts. In this respect it is noteworthy that *B. thuringiensis* is closely related to *B. cereus*, a ubiquitous saprophyte that lacks the ability to produce δ-endotoxin.

Bacillus popilliae

The bacteria grouped under the name *B. popilliae* cause *milky diseases* of beetles (Coleoptera), specifically of the beetle family Scarabaeidae. This family includes the beneficial dung beetles but also some of the most important pasture pests – the chafers. In practice *B. popilliae* has been used intensively and almost exclusively for control of the Japanese beetle *Popillia japonica* in the USA, and to a lesser degree against the European corn chafer *Amphimallon majalis* in that country. In these highly restricted uses and in some other respects it provides interesting contrasts with *B. thuringiensis*. For example it is highly pathogenic in its own right and does not need toxins to kill its hosts (though toxins may be involved in disease development). Also it is highly persistent in the environment so it can be used for mass release to achieve lasting control. But, unlike *B. thuringiensis*, the spores of *B. popilliae* cannot be produced in artificial media, so the inoculum for control programmes must be produced in living hosts.

The story of *B. popilliae* centres around the Japanese beetle which was accidentally introduced into the USA from Japan early this century. Although not a problem in its area of origin, the beetle began to cause serious damage in the USA; it spread rapidly from the initial sightings in New Jersey (1916) and today it is found over roughly half of the country, in almost every state east of the Mississippi. It is a problem as an adult beetle because it feeds on a wide range of plants, eating out the leaf tissues between the leaf veins, and it accumulates on ripening fruit causing substantial damage. In addition it is a problem in the larval stage because the adult beetles lay their eggs in grass turf and the grubs destroy the grass roots, especially on new housing estates where natural enemies are absent. By the 1930s the beetle problem had become so serious that a search was begun for a control measure and this led to the discovery of some naturally occurring diseased larvae. The disease was termed 'milky disease' because of the milky white appearance of the grubs, due to a large number of refractile bacterial spores in the haemolymph (Figure 3). Two types of bacterium were subsequently isolated from two types of milky disease. Type A disease was characterized by a pure white appearance of the grubs and the bacterium in this case was named *B. popilliae*. Type B disease differed in that the grubs showed a transition from white to bjown over winter and the bacterium causing this disease was named *B. lentimorbus*. More recently a range of other milky disease bacteria have been isolated from beetle hosts throughout the world, and two other species have been formally described – *B. fribourgensis* and *B. euloomarahae*.

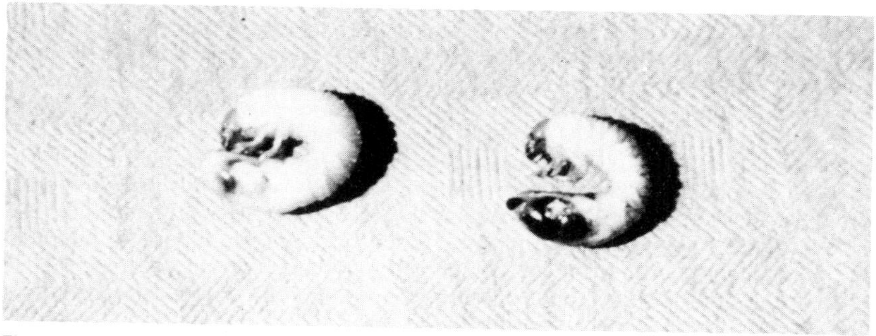

Fig. 3 Grubs of the Japanese beetle, *Popillia japonica*. Healthy grub, right; milky diseased grub, left. The grubs are 2 to 3 cm long.

The taxonomic position is now changing, the trend being to regard all of these as varieties of *B. popilliae*. For example, *B. lentimorbus* is now called *B. popilliae* var. *lentimorbus*. The change from species to varietal rank reflects the view that these bacteria are more closely related to one another than they are to other *Bacillus* spp. The important point, however, is that the varieties differ in host range – at least in field conditions – and thus in their potential for microbial control of different pests.

The bacterium and its physiology *B. popilliae* is a Gram-negative spore-forming rod, 1.3 to 5.2 × 0.5 to 0.8 μm, but it sometimes becomes Gram-positive at the start of sporulation. It is a fastidious organism that grows only on rich media containing yeast extract, casein hydrolysate or an equivalent amino acid source, and sugars. Several amino acids are known to be required for growth, as are the vitamins thiamine and barbituric acid. Trehalose, the sugar found in insect haemolymph, is a favoured carbon source though glucose also can be used. It is significant that trehalose is taken up exclusively by the phosphoenolpyruvate (PEP) : sugar phosphotransferase system which requires an adequate supply of PEP.

Some varieties of *B. popilliae* form a parasporal crystalline body inside the cell at the time of sporulation and in this respect resemble *B. thuringiensis*. But the crystal is not thought to play a significant role in infection and certainly it is not as important as in *B. thuringiensis*. The variety *lentimorbus*, for example, does not produce a crystal and yet it causes disease. Another difference between *B. popilliae* and *B. thuringiensis* is that *B. popilliae* cannot be induced to sporulate in laboratory media although it does so readily in the diseased host. Actually there are a number of oligosporogenic mutants – ones that produce a few spores – but they have no commercial value because the spores are infective only on injection into insects, not by the normal route of ingestion. For this reason, spores for microbial control programmes must be produced in living insect larvae – an expensive and time-consuming process.

The cellular control of sporulation has been studied intensively with a view to simplifying spore production (Bulla *et al.*, 1978). Whereas most *Bacillus* spp. sporulate readily at the end of the exponential growth phase in culture, *B. popilliae* does not sporulate and instead its cells rapidly lose viability. This loss of viability has been explained in terms of the failure of *B. popilliae* to produce catalase and peroxidase enzymes. Ordinarily, these prevent the accumulation of hydrogen peroxide which is toxic and formed in the normal course of metabolism, and it is significant that the oligosporogenic mutants (which do not lose viability at the end of exponential growth) do produce a catalase at the onset of sporulation. This does not explain why wild-type cells cannot sporulate in artificial media but do so in the host. The suggested reason for this centres around sugar metabolism. Insect haemolymph is known to contain high concentrations of trehalose even at the onset of sporulation, and yet similar levels of trehalose completely suppress sporulation by the oligosporogenic strains in artificial media. So it is argued that the internal rather than the external sugar concentration may be critical for sporulation, and this of course can be regulated by membrane transport processes. As already noted, intracellular levels of PEP regulate the uptake of trehalose by the PEP : sugar phosphotransferase system. PEP is an intermediate of the Embden-Meyerhof pathway of glycolysis but not of the alternative pentose-phosphate pathway, and the evidence suggests that *B. popilliae* uses the pentose-phosphate pathway when growing inside the insect but the Embden-Meyerhof pathway in laboratory culture. Summarizing, it is suggested that the different pathways of sugar metabolism *in vivo* and *in vitro* result in different intracellular levels of PEP; these in turn determine the rate at which trehalose is taken up from the surroundings, and trehalose at too high a level inside the cell represses sporulation.

The host-parasite interaction *B. popilliae* causes disease of beetle larvae when they ingest spores in the soil. The spores germinate in the gut within 2 days and the vegetative cells proliferate, attaining maximum numbers within 3 to 5 days. By this time some of the cells have penetrated the gut wall and begun to grow in the haemolymph, where large numbers of cells develop by day 5 to 10. A few spores also are formed at this stage but in the variety *popilliae* the main phase of sporulation occurs later and is completed by 14 to 21 days when the larva develops the typical milky appearance. In laboratory conditions the larva remains alive until this stage and usually contains about 5×10^9 spores. In field conditions, however, there are reports that larvae sometimes die earlier, before the main phase of sporulation is completed. This is of concern because sporulation stops when the host dies and the larva ultimately releases fewer spores to maintain the level of infestation of a site.

The cause of insect death is not fully known. Physiological starvation caused simply by the number of bacterial cells seems the most likely explanation, and fat reserves of diseased larvae have been shown to be much reduced compared with those of healthy larvae. However, toxins also may be involved because they have been detected in culture filtrates of the bacteria and shown to be lethal on injection.

Host ranges of the milky disease bacteria are difficult to assess because of the problems of obtaining different hosts at the same developmental stage and of obtaining equivalent batches of inoculum. The available results are intriguing because they suggest that any single variety of *B. popilliae* causes disease of a wide range of beetle hosts on injection, of a much narrower range of hosts by the normal route of ingestion, and of a very narrow range of hosts in normal field conditions. The reasons for such specificity in nature are largely unknown.

Application for microbial control *B. popilliae* has been registered for control of the Japanese beetle in the USA since about 1950 – the first registration of any insect pathogen as a microbial control agent. The control strategy is aimed solely against the larvae, so if the beetle itself is causing serious damage a chemical insecticide must be used for short-term control. The bacterial spores are produced commercially in larvae collected from grass turf on golf-courses, airports, etc. The larvae are injected with cells, incubated until they develop a milky appearance and then crushed and dried to give a spore powder. In 1981 this was done by two firms in the USA: by Fairfax Biological Laboratory Inc. who marketed the products 'Doom' and 'Japidemic' and by Reuter Laboratories Inc. who marketed 'Milky Spore'. Spore powders are applied to turf in small heaps at roughly 1-metre spacing and the spores are then distributed naturally by wind and rain. They can persist in soil for several years and therefore give lasting control of a pest problem, the spore numbers being boosted periodically when a diseased larva dies.

The use of *B. popilliae* has proved remarkably successful. Between 1939 and 1953 over 100 tons of spore powder were applied to turf in over 160,000 sites in the USA as part of a Government programme (Fleming, 1968). Larval numbers in the turf were reduced 10- to 20-fold and the population stabilized at this new low level, with corresponding reductions in the levels of adult beetle damage. However, the treatment is effective only when applied on a region- or state-wide basis (or at least to relatively large areas) to reduce overall the levels of beetle infestation. It is least appropriate for use by small landowners, who may control the larvae in their own turf only to find their trees and shrubs being eaten by beetles from their neighbours' properties. Also, because *B. popilliae* is obligately dependent on its hosts for sporulation and because some larvae may not ingest spores (or not ingest enough to cause disease) a periodic resurgence and decline of the pest problem can be expected as shown in Figure 1. The success of the control programme must be judged not on this basis but by the fact that over a number of years the mean level of pest damage is lower than it would be in the absence of *B. popilliae*.

Advantages and disadvantages of *B. popilliae* The advantages of *B. popilliae* include (1) its very narrow host range (which is environmentally desirable) and its consequent lack of effect on beneficial insects; (2) its complete safety for man and other vertebrates (for example, it does not grow at 37°C); (3) its compatibility with other control agents including chemical insecticides; (4) its persistence, giving lasting control. Its disadvantages, however, include (1) the high cost of production *in vivo*; (2) its slow rate of action; (3) most importantly, its lack of effect on adult beetles which often cause the most obvious and distressing damage, and (4) its relative unattractiveness to the small landowner. It is significant that the narrow host range and consequently restricted market for the product, coupled with the difficulty of applying large-scale methods for spore production, have prevented the major agrochemical industries from showing interest in producing inoculum.

Outstanding problems and future research Clearly, more work is needed on the sporulation of *B. popilliae* with a view to producing spores commercially in culture. This might not have a major impact on its use against the Japanese beetle – a use that is already widespread and successful in the USA – but it could enable the bacterium to be used against beetle pests of lesser status for which the high cost of current spore inocula cannot be justified. More work is also needed on the way in which *B. popilliae* kills its hosts and why it sometimes causes early death before a full complement of spores has been produced. If death is due to a toxin this might be removed by strain selection. The result would be less dramatic kills in the short term but more effective and persistent control in the long term.

Of particular concern is a report of the re-emergence of the Japanese beetle as a serious pest in Connecticut, in regions where it had been controlled effectively since the initial applications of spore dust in the 1940s (Dunbar & Beard, 1975). Larval densities ranged from 0 to 474 per square metre of turf in 1974 (mean 112), and were sometimes as high as those recorded 25 years earlier, before the control programme was begun. Moreover, in this study only 0.2% of larvae collected from field sites showed symptoms of milky disease compared with 41.5% disease incidence in a survey in 1946 after *B. popilliae* had been introduced. Spores collected from these few diseased larvae caused only 7 to 17% infection of larvae in laboratory tests, compared with 65 to 67% infection from spores collected from New York State where a decline in the degree of control had not been reported. Even this figure was low in relation to the expected 90% disease incidence at the inoculum level used. Clearly there has been a marked reduction in virulence of *B. popilliae* in Connecticut over the years and a corresponding resurgence of the pest problem. Perhaps the bacterium is losing virulence also in New York State but not yet to a degree that limits its effectiveness. Detailed study of this is required, as also of the possibility that the pest itself has developed some resistance to the disease.

Little is known about the effects of mixed inoculations of virulent and attenuated strains of pathogens into insects, nor about the physiological basis of any resistance induced by attenuated strains. This might have a bearing on any attempts to reintroduce fully virulent strains where the control method has broken down. Lastly, all these lines of investigation depend on being able accurately to assess the virulence of a strain and to compare it with others. This in turn requires further work on the effects of long-term storage on virulence, so that comparisons can be made over periods of time.

Summary: comparison of *B. thuringiensis* and *B. popilliae*

Both *B. thuringiensis* and *B. popilliae* have been used successfully as microbial control agents over two decades. Their different characteristics are summarized in Table 2.

B. thuringiensis affects nearly 300 lepidopterous pests, namely those with a mid-gut pH more than 8.9. It is marketed as a mixture of spores and crystals, the latter being broken down in the gut of susceptible larvae to release δ-endotoxin. This causes gut paralysis in all affected species, and in some it leads rapidly to a general paralysis. Either the toxin alone is lethal or its effects are followed by bacterial invasion resulting in a lethal septicaemia. In the absence of the toxin the bacterium is only weakly pathogenic. The spore and crystal mixture is produced commercially *in vitro*. It is applied to plants either as a dust or as a liquid suspension to give rapid but usually only short-term control. Several commercial preparations are available and they are registered for use on food crops right up until harvest time if necessary.

Table 2 Comparison of the properties of *B.thuringiensis* and *B.popilliae* as microbial control agents

	B. thuringiensis	*B. popilliae*
Pests affected	Lepidoptera (many)	Coleoptera (few)
Pathogenicity	Low	High
Speed of effect	Immediate	Slow
Role of toxins	Important	Questionable
Current use	Widespread	Restricted (USA)
Method of use	Microbial insecticide	Introduction
Formulation	Spores + toxin crystals	Spores
Production	*In vitro*	*In vivo*
Persistence	Little or none	Marked
Resistance in pest	None reported	Reported but unconfirmed

B. popilliae affects only scarabaeid beetles and has been used almost exclusively for control of the Japanese beetle in the USA. It is marketed as a spore dust for application to grass turf and it causes disease of larvae as a direct result of its aggressive pathogenicity. The spores can be produced only *in vivo* and in 1981 this was done only by two small companies. The spores are used to achieve lasting control because they are highly persistent and the spore population is replenished periodically when diseased larvae die. The bacterium is relatively slow-acting, however, both in individual larvae and in the population as a whole, and it is applied most effectively on a region- or state-wide basis to reduce the mean level of beetle damage.

Both control agents cause disease on ingestion; both affect only (or predominantly) the larval stages of their hosts; both are harmless to man and other vertebrates and also to the majority of beneficial insects; both are largely unaffected by chemical insecticides; and both have a reasonable 'shelf-life' during storage. In both cases, moreover, the microbiologist has played a significant part in the development of the microbial control programme and has a major part to play in the future.

The prospects for using other bacterial pathogens of insects are less encouraging. Very few of them are specific for insect hosts, most are at best only weakly pathogenic on ingestion though they can kill insects on injection, and most do not produce spores and therefore might not survive long in the field. *Bacillus cereus* is one of the commonest bacteria isolated from diseased insects in nature and it can be highly pathogenic, presumably as a result of its phospholipase activity which may help it to invade through the gut wall. However it is taxonomically very close to *B. thuringiensis* and it lacks the specific attributes that have made *B. thuringiensis* so successful as a microbial control agent, namely a selective mode of action and yet relatively broad host range amongst insect pests.

References

BULLA, L. A., COSTILOW, R. N. and SHARPE, E. S. (1978). Biology of *Bacillus popilliae*. *Advances in Microbial Physiology* 23: 1–18.

CANTWELL, G. E. (Ed.) (1974). *Insect Diseases. Volumes I and II.* Marcel Decker Inc., New York.

DUNBAR, D. M. and BEARD, R. L. (1975). Present status of milky disease of Japanese and Oriental beetles in Connecticut. *Journal of Economic Entomology* 68: 453–7.

ERMAKOVA, L. M., GALUSHKA, F. P., STRONGIN, A. YA., SLADKOVA, I. A., REBENTISH, B. A., ANDREEVA, M. V. and STEPANOV, V. M. (1978). Plasmids of crystal-forming bacilli and the influence of growth medium composition on their appearance. *Journal of General Microbiology* 107: 169–71.

FALCON, L. A. (1971). Use of bacteria for microbial control of insects. In *Microbial Control of Insects and Mites* (Ed. H. D. Burges and N. W. Hussey). Academic Press, London.

FAST, P. G. (1981). The crystal toxin of *Bacillus thuringiensis*. In *Microbial Control of Pests and Plant Diseases 1970–1980* (Ed. H. D. Burges). Academic Press, London.

FLEMING, W. E. (1968). Biological control of the Japanese beetle. *United States Department of Agriculture Technical Bulletin Number 1383.* Washington DC.

3 Microbial control of pests: use of viruses

Viruses have great potential as insect control agents but their use is limited at present by concern about their safety. The first registration of a viral product in the USA was in 1975 for control of the cotton bollworm, and still very few are used commercially. Nevertheless viruses frequently cause population crashes of their hosts and thus are exploited indirectly. All those currently used or seriously being considered for commercial use belong to one family, the *baculoviruses* or Baculoviridae. These have very characteristic morphologies and seem to be entirely restricted to insect hosts; indeed even within the class Insecta most of them have very restricted host ranges. Their most successful applications have been by 'introductions' to give long-term control but some also have been used for season-long control in annual crops.

There are mammalian pests as well as insect pests, and one of the most spectacular of all biological control programmes concerned the use of myxoma virus which decimated the rabbit populations in Australia and Europe in the 1950s. The disease *myxomatosis* caused a huge public outcry and did much to diminish the standing of scientists in the community at that time. It is treated separately at the end of this chapter because it can teach us a lot about the practice of microbial control.

The types of insect virus

There are six main groups of insect virus as shown in Table 3. Some of them closely resemble the viruses of vertebrates and thus are automatically excluded from consideration as potential control agents; although they probably do not replicate in warm-blooded animals the proof of their safety would be prohibitively expensive. Others are clearly different from any viruses recorded from vertebrates and they include three types: the *nuclear polyhedrosis viruses* (NPV), the *granulosis viruses* (GV), and the *cytoplasmic polyhedrosis viruses* (CPV). These are shown in Figures 4 and 5, and their main features as potential control agents are outlined below. At this stage it can be said that the NPV and GV are so similar that they are included in one family, the baculoviruses.

All three types (NPV, GV and CPV) are *occluded viruses*; that is, the virus particles or virions are enclosed in a proteinaceous shell which has a paracrystalline structure and is termed an *inclusion body*. This gives a measure of protection that the individual virions would not have and it is an important feature in their use as microbial control agents. Moreover, the fact that no such inclusion body has ever been seen in mammalian cells lends support to the view that these viruses present no real environmental hazard.

In the NPV (Figure 4) the virions are rod-shaped and each is surrounded by an envelope. The enveloped virions usually occur in groups surrounded by a membrane, and several such groups are contained within the inclusion body. As the name implies, the inclusion body is polyhedral, ranging from 0.5 to 15.0 μm

Table 3 Characteristics of insect viruses

Family	Common name or representative genus	Particles				Nucleic acid		Site of replication
		Shape	Size (nm)	Occluded	Enveloped	Type	Mol. wt ($\times 10^6$)	
Baculoviridae	*Baculovirus* (NPV and GV)	Bacilliform	250–400 × 40–70	+	+	ds DNA	50–100	Nucleus
Iridoviridae	*Iridovirus*	Isometric	130–200	0	some	ds DNA	{114–150 240–288	Cytoplasm
Poxviridae	Pox virus	Complex multi-layered	up to 300–450	+	+	ds DNA	123–200	Cytoplasm
Parvoviridae	*Densovirus*	Isometric	20–22	0	0	ss DNA	1.6–2.2	Nucleus
Reoviridae	includes CPV	Isometric	50–65	+	0	ds RNA	about 15	Cytoplasm
Picornaviridae	*Enterovirus*	Isometric	27–30	0	0	ss RNA	2.6–3.2	Cytoplasm

diameter, and the virus replicates in the host cell nuclei. Because of their size the polyhedra are readily seen under a light microscope.

The GV are similar to the NPV in all respects except that the inclusion bodies are smaller (0.2 to 0.5 μm diameter), are ovoid or elliptical and usually contain only one virion. They appear granular under the light microscope – hence the name granulosis virus.

The CPV differ from the others in several respects. The virions are isometric and lack an envelope. They are enclosed in polyhedral inclusion bodies which resemble those of NPVs but the polyhedra are relatively unstable and virions near the surface can be removed by mild chemical treatment (Figure 5). As the name CPV implies, replication occurs in the cytoplasm rather than the nucleus of host cells.

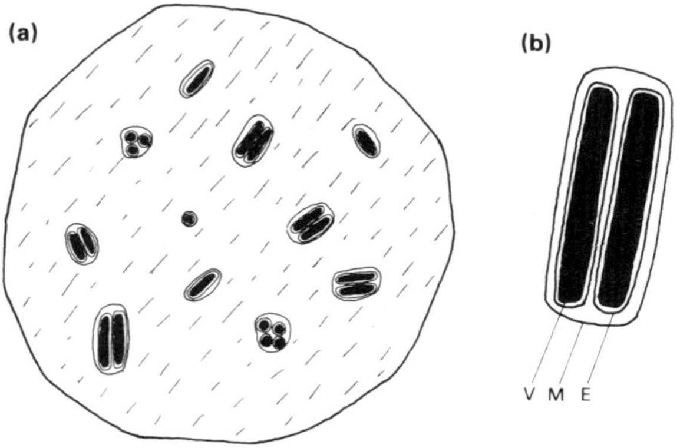

Fig. 4 Nuclear polyhedrosis virus. **a** Polyhedron with paracrystalline structure enclosing random groups of viruses surrounded by a membrane. **b** Two virus particles (V) each surround by an envelope (E) and enclosed in a membrane (M).

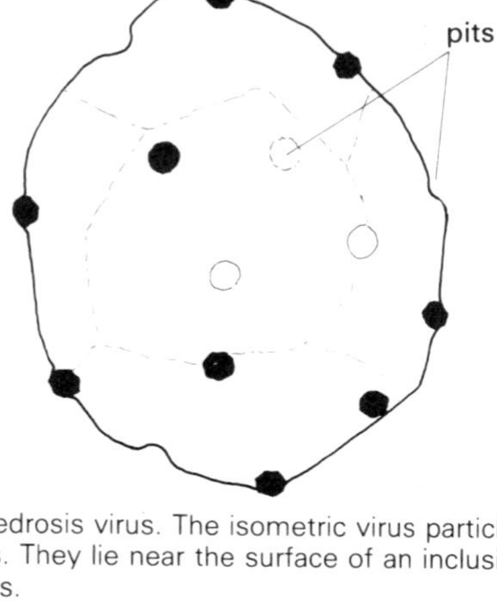

Fig. 5 Cytoplasmic polyhedrosis virus. The isometric virus particles are not enclosed in envelopes or membranes. They lie near the surface of an inclusion body and are lost readily from it, leaving pits.

Host range and mode of infection

About 125 types of NPV have been described, of which roughly 90% occur in the Lepidoptera (butterflies and moths) and others occur in the Hymenoptera (especially sawflies), Diptera (cranefly and mosquito) and Orthoptera (grasshoppers, locusts). The precise host ranges are unknown in all cases, owing to the high frequency of inapparent infections caused by insect viruses. Nevertheless, in general the NPV seem to be family-specific and there is probably no cross-infection between, for example, Lepidoptera and Hymenoptera. About 50 GV have been recorded, again mainly from the Lepidoptera. In general these viruses show greater host-specificity than do the NPV and in some cases they may be genus- or even species-specific. The CPV have been reported from about 200 insects, also mainly in the Lepidoptera, but they are the least host-specific of the viruses discussed here and there is a consequently greater chance of their proving hazardous – unlikely though this is.

In all three types of virus, infection occurs via the gut when inclusion bodies are ingested by insects feeding on contaminated vegetation. The inclusion body dissolves in the mid-gut, releasing virions which penetrate the lining epithelium and start to replicate. The subsequent course of infection differs between virus types and also between types of host. In the Lepidoptera the NPV and GV normally have only a very restricted phase of development in the gut epithelium; they seldom form inclusion bodies in these cells and instead they produce virions which pass from the gut epithelium into the tracheal epithelium and then into the haemolymph and other tissues. Typical inclusion bodies are found in all the subsequently invaded cells, but throughout the course of infection some free virions are produced and they presumably serve to spread the infection within the host. The time taken to kill the insect depends on the virus dose, the stage of development of the host and various other factors like the degree of stress of the insect; it can range from 6 to 24 days or more. As an example, Lewis (1981) records LT_{50} values (the times taken for 50% of challenged insects to die) of 5 to 12 days in the case of an NPV in larvae of the gypsy moth *Lymantria dispar*. The LD_{50} values in this case (i.e. doses needed for 50% mortality) ranged from about 200 to 1700 inclusion bodies per larva, depending on the batch of host larvae tested. The dead insect usually contains a large number of inclusion bodies – with NPV yields of 6 to 43 $\times$ 10^9 polyhedra per gram of larva are not uncommon – and these are finally released by rupture of the insect's integument.

Infection of sawflies (Hymenoptera) by NPV differs somewhat from the pattern just described. Usually the gut epithelium is very heavily colonized and new inclusion bodies are formed in it. They are released into the gut and either regurgitated or voided with the faeces, so fresh inoculum is released from the diseased insect throughout the course of infection. This may help to explain why the baculoviruses have proved to be such spectacular control agents of sawflies (see later). Infection of insects by CPV is often much more protracted than that by NPV and GV, but the CPV eventually colonize and destroy nearly all of the gut epithelium, releasing polyhedra which are voided with the faeces. Numerous references to the details of these infection processes can be found in the review by Tinsley (1979).

Viruses can infect nearly all developmental stages of their hosts, though the early larval stages are usually the most susceptible. If infection occurs late in larval development or if the progress of infection is slow because of an initially low virus dose, the larva may pupate and give rise to an infected adult. This in turn can produce contaminated eggs and thereby pass the infection to the next generation. It is unclear how many of these infections occur inside the eggs and how many result from surface contamination of the eggs. The latter has been demonstrated many times and has been suggested as a possible basis for practical control programmes by applying virus paste to the genital regions of adult females and then releasing them into the field. Insects that parasitize other insects have also been shown to transmit viruses from one larva to another in laboratory conditions (by injection during egg-laying) but the significance of this in field conditions is unknown.

Applications for microbial control

The CPV are generally less appropriate than the NPV and GV as insect control agents because of their relatively wide host ranges, the slow progress of disease and the relative instability of their polyhedra. Apparently the only one to be registered as a control agent by 1980 was a CPV of the pine caterpillar *Dendrolimus spectabilis* in Japan 1974, though others are doubtless exploited indirectly as natural mortality agents. The rest of this section will concentrate on the baculoviruses.

The most spectacular successes to date have been by means of 'introductions', especially for forest pests. For example, the European spruce sawfly *Gilpinia hercyniae* (Hymenoptera) became a serious problem in Canadian forests in the 1930s following its accidental introduction from Europe. In 1931 an aerial survey showed that spruce was subject to severe defoliation in over 2,000 square miles and the pest was beginning to spread rapidly from the originally identified outbreak area. A major programme was begun in which natural parasitic insects were imported from Europe and mass-reared for release into the forests. Initially the level of release was low, but 2.5 million parasites were released in 1935, over 18 million in 1936, over 47 million in 1937 and 222 million in 1940. Meanwhile the sawfly problem continued to get worse; in 1939, 73% of white spruce and 43% of black spruce in the original outbreak area were reported dead (a bark beetle contributed to the losses) and the infestation was spreading further afield than ever before. By 1942, however, the larval populations in all but a few areas had declined to medium or light intensity and this was associated with a viral disease caused by an NPV. The origin of the virus has never been traced but it was probably introduced unknowingly with one of the parasites from Europe. The virus, together with the parasitic insects, has contained the pest problem ever since. A rather similar situation occurred in Britain when this sawfly was accidentally introduced there from its indigenous region in Central and Northern Europe. The pest initially caused serious damage to spruce plantations but an NPV that was already present in Britain caused a natural epizootic and largely controlled the problem. The original host of this virus has not been identified. One interesting feature of this example is the evidence that has been obtained to implicate birds as dispersal agents of the virus. Entwistle *et al*. (1977) found that 80% of the total bird number surveyed in an area of spruce forest in mid-Wales dispersed the virus in their droppings, having presumably eaten infected sawfly larvae; the virus was shown to be infective after passage through the gut.

Another similar example concerns the European pine sawfly *Neodiprion sertifer* which caused serious damage to pines in parts of the USA in 1943, following its accidental introduction from Europe. In this case, however, the potential value of viruses was recognized at an early stage and a few larvae naturally infected with an NPV were introduced from Sweden. The oral LD_{50} of the virus was calculated at between 100 and 500 polyhedra per larva, and field trials showed that sprays containing 10^6 polyhedra/ml gave rapid and high mortality whilst 10^4/ml gave similar mortality but more slowly. The latter was adopted as being most economical and was used routinely to give short-term protection to young trees. In fact the virus soon became generally distributed and gave a large measure of natural control, but young trees are particularly susceptible to damage and need the extra protection provided by a spray. This virus has been marketed by a company in Finland since 1972; the treatment cost in 1978 was about $US 11/hectare, which is relatively expensive because of the difficulty of rearing sawfly larvae for inoculum production.

The first commercial viral pesticide for use in agricultural crops was marketed in 1975 under the trade name 'Elcar' (Sandoz Inc., USA); it is an NPV for control of the cotton bollworm *Heliothis zea*. The virus infects 7 species of *Heliothis* – a genus that includes some extremely important crop pests – but not 37 other tested insects, spiders or mites. It is produced in artificially reared larvae in a pilot plant with a current input of 10^6 larvae per month. The larvae are fed contaminated diet, incubated until they are dead or dying and then left to putrefy in water when the inclusion bodies sink as a precipitate. The recommended rates of use on cotton in the field are 10×10^{10} polyhedral inclusion bodies per acre (0.4 hectare) for light or moderate pest levels and up to 45×10^{10} polyhedra per acre for heavy infestations, with an average rate of 24×10^{10} which represents 25 larval equivalents and a cost of $3.12 per acre. It is hoped that in the near future the use of this virus will be extended to cover control of *Heliothis* spp. on soybeans and other food crops (Ignoffo & Couch, 1981). A significant point raised by this example is the need for a large supply of insect larvae to ensure efficient production of inoculum. This became feasible only when research in the 1960s led to the development of artificial diets for a range of insect species so that the hosts could be reared at times of the year when their natural foods were unavailable. There is still no satisfactory artificial diet for sawflies and this is one of the contributory factors to the high cost of viruses for control of these pests (see earlier).

One last example of the use of viruses as control agents will show how in practice the release of virus-infected adults can be effective. The rhinoceros beetle *Oryctes rhinoceros* has been spreading into the Pacific Islands from its endemic regions in south-eastern Asia, causing serious losses in coconut palms and oil palms (Bedford, 1980). The adult beetles feed on the leaves and bore into the growing points of the plants where they cause damage themselves and also create wounds through which decay fungi can enter. The larval stages are found in dead palms, sawdust heaps and many other types of decaying organic matter, with the result that chemical control measures are unfeasible and even sanitation is not fully effective. In 1963 a baculovirus was discovered in parts of the beetle's natural range and attempts were then made to introduce it into the Pacific Islands. After trying various methods the most practicable was found to be by releasing contaminated adult beetles that had been immersed in a viral suspension for 2 to 3 minutes and then allowed to crawl in contaminated sawdust for 24 hours. The virus is most damaging to larvae but it can multiply in the gut epithelium of the adults and is then released into the gut and voided with the faeces. In field conditions the infected adults live for a considerable time and spread the infection both to other adults and to the breeding sites. As shown in Figure 6, in the Fiji Islands this control measure has reduced the amount of damage to palm fronds from initial values of between 30 and 90% to final values of about 5 to 20%, and this has been achieved at very low cost. The significant point about the example is that the control measure makes use of the insect's own habitat-selection behaviour to introduce the virus into the sites where it is likely to be most effective. Incidentally, the example is interesting in another respect because the virus in this case is not usually occluded, and certainly not in polyhedra, although its other features place it in the baculovirus group.

Advantages and disadvantages of the use of viruses

In many respects the same considerations apply to viruses as to *Bacillus popilliae* in the previous chapter. Obligate parasites can be produced only *in vivo* and therefore at some inconvenience and considerable expense, though in the case of viruses the yields from each diseased larva are high enough to make this economical. Also, the effects of viruses are not immediate, and control can take some time to be achieved if an insect ingests too few virus particles. Short-term chemical control may therefore be needed, especially for crops whose market value is affected by even minor damage. Furthermore, viruses act only against the surface-feeding stages of their hosts because the virus particles are confined to the plant surface, and they are inactivated quite rapidly by UV radiation which is one of the most serious practical problems in the field. Their advantages include the persistence of baculoviruses in the soil environment, because the inclusion bodies are resistant to bacterial lytic enzymes, which enables them to bring about lasting control of some pests. Also they cause minimal disruption of the natural enemy complex and as far as is known they pose no threat to man or other vertebrates. As

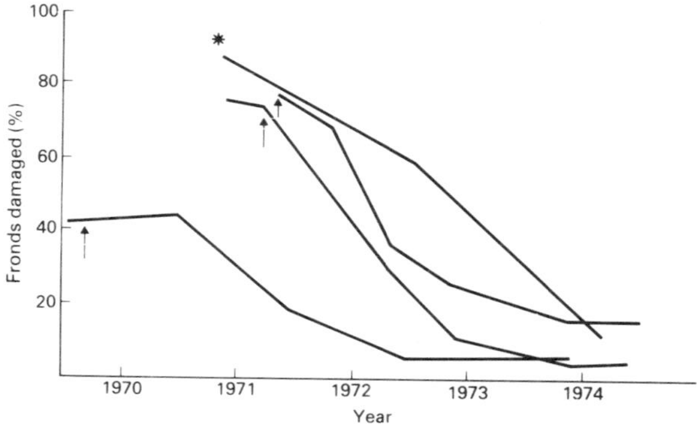

Fig. 6 Effect of a baculovirus on palm damage by *Oryctes rhinoceros*, the rhinoceros beetle, in localities in the Fiji Islands. In three localities the time of virus introduction is marked by an arrow; in the fourth (*) the virus spread naturally into the area from another site of introduction. (Taken from Bedford (1980) by kind permission of the author and publisher.)

with *B. popilliae*, introduction of a virus does not eliminate a pest and there can be periodic resurgences of a pest problem followed by equally striking reductions (Figure 1). If necessary, as we saw in the case of the pine sawfly, virus can be sprayed on to a crop at a particularly vulnerable stage of its growth to enhance the degree of protection beyond that provided by the background population of the virus.

Outstanding problems and future research

Undoubtedly the main problems with insect viruses centre around safety considerations. This topic is covered in depth in a book edited by Summers *et al.* (1975). All available evidence suggests that the baculoviruses do not infect vertebrates, and yet caution is justified because micro-organisms, unlike chemicals, can replicate themselves and it is possible that any mistake at this stage could not be rectified. All authorities agree that a major limitation to declaring the baculoviruses absolutely safe is the high and unexplained frequency of inapparent infections caused by these and other insect viruses in insect hosts. In practice this makes it difficult to devise adequate protocols for safety testing because absence of symptoms does not necessarily mean absence of infection. Inapparent infections of insects can become overt when the host is stressed by environmental conditions or when challenged by inoculation with another pathogen – even one that does not normally infect that host. Clearly, much more research is needed into inapparent infections and their physiological bases.

In a different context but still in the realm of insect pathology, there are several reports of interference and antagonism between viruses in mixed inoculations, and also some reports of synergism. The mechanisms in these cases are almost completely unknown but this subject has great potential significance for microbial control. For example, what effect is a background population of viruses likely to have when a new virus is used as a control agent?

Lastly, there is increasing interest in the possibility of using cultured insect cell lines to produce viruses commercially for control programmes. This would not only 'streamline' production of inoculum but it might also be essential in the future to meet the strict requirements of registering authorities in terms of purity of viral preparations (Summers *et al.*, 1975). Typical inclusion bodies have been produced on an experimental scale in cultured cell lines of the silkworm, the cotton bollworm and the cabbage white butterfly, among others, especially if haemolymph from diseased insects is used as the inoculum. In theory the viral yields should be as high as those from intact insects but in practice they are much lower owing to technical difficulties. An additional problem is that the media on which the cell lines are grown are complex and expensive, because animal cells tend to lose essential nutrients to the surrounding medium – even those that they have synthesized themselves – and these nutrients must therefore be supplied at high enough levels in the surrounding medium to prevent their loss by diffusion from the cells. Even if these difficulties could be overcome there remain some microbiological problems, like the fact that the production of GPV in cell lines has proved intractable, although the NPV can be grown quite easily.

Myxomatosis

The European rabbit *Oryctolagus cuniculus* was once a serious pest in Europe and Australia. It is estimated to have caused about 6% yield loss of winter wheat in Britain; it damaged young trees in parks and woodlands, and sometimes it grazed natural pastures so low that it reduced their carrying capacity for livestock. By grazing pastures it prevented the flowering of many rare plants on chalk downlands and sand dunes, and its burrowing activities were an important factor leading to the erosion of stabilized sand dunes by wind. The huge reduction in rabbit numbers that followed the outbreak of myxomatosis in the 1950s was thus welcomed by farmers, foresters and conservationists alike. But it was not without cost. Some of the former pastures rapidly became scrub and woodland because the grazing pressure of rabbits had prevented woody plants from becoming established; this in turn led to the permanent destruction of the habitats of some rare plants. The traditional role of the gamekeeper changed and some were made redundant. The hunting and fur trades were seriously hit; indeed, a court action was brought against Dr Armand Delille, the French landowner who first introduced the disease into Europe (the action failed on a technicality). Above all, the public was horrified that scientists were directly or indirectly associated with the introduction of a tumorigenic virus into a population of vertebrates. The disease itself is horrifying because rabbits become blind and deaf, wander aimlessly and sometimes develop weeping tumours before they die. 'Mercy squads' roamed the countryside with sticks to kill affected rabbits. In short, myxomatosis holds a place in both the vegetational and social history of Britain, though it should be stressed that the disease was not intentionally started here and efforts were made unsuccessfully to contain it.

The story of myxomatosis begins in 1897 when Sanarelli, a medical researcher in Uruguay, almost completely lost his stock of European rabbits as a result of an unknown disease. Their faces were swollen, especially around the eyes, and their bodies were covered with small tumours that released a sticky mucinous fluid when cut. He named the disease myxomatosis (Gr. *myxa* = slime) and suggested that it was caused by a filterable agent. This is now called the *myxoma* virus; it is a member of the pox group, 290 × 230 nm, with double-stranded DNA and a molecular weight of 130 to 240 × 10^6. Over the next half-century there were further reports of sporadic outbreaks of the disease in South America, but little progress was made until 1942 when Aragao showed that the native Brazilian rabbit *Sylvilagus brasiliensis* is immune to infection. This was shown to be acquired immunity, suggesting that the virus is endemic in South America even though the disease is seldom seen there.

The prospect of using the virus as a control agent was recognized in Australia, and during the 1930s and 1940s the Australian government commissioned research in both Australia and Europe to investigate this possibility. Full details of this phase of the study are recorded by Fenner & Ratcliffe (1965). Laboratory and cage trials were highly successful, but results from the field trials were disappointing and the work ended with the conclusion, expressed by one research group, that 'myxomatosis cannot be used to control rabbit populations under most natural conditions in Australia with any promise of success'. By 1949, however, the rabbit problem in Australia had become so serious that further study seemed justified, and this was reinforced by the view of some scientists that the original field trials had been unsuccessful because they were done in geographically isolated areas (for quarantine purposes). In 1950 the disease was introduced into the Murray River valley, a moist region of New South Wales. At first this also seemed to have failed, but late that year spontaneous outbreaks were reported from numerous parts of the region, sometimes quite far from the sites of introduction. In the next few weeks the disease was reported over much of south-eastern Australia and wherever it occurred the rabbit population was severely reduced. The most complete records are for two observation areas at Lake Urana, NSW, where rabbit numbers were reduced from 5,000 to 50 and from 500 to 12 in 1951–2. Tests on survivors showed that only 20 to 30% had antibodies to the virus and therefore had been exposed to infection. So on this basis the case-mortality rates were calculated at 99.8 and 99.4% respectively. Death occurs rapidly after exposure; there is an interval of 5 to 7 days before the symptoms appear but the animal then often dies 2 to 4 days later.

The outstanding success of this control programme was found to depend on the efficiency of the mosquito as a vector, and this at least partly explains why the earlier introductions in isolated areas were unsuccessful. Man also played a part in distributing the disease because, despite legal restrictions, farmers were known to travel large distances in search of affected rabbits to introduce on to their lands. By this time the virus was known to pose no threat to man and his domesticated animals; apart from rabbits only a few individuals within two species of hare are severely affected by it.

The first myxomatosis outbreak in Europe is equally well documented (Greathead, 1976). It was started deliberately in France in 1952 by Dr Delille on his walled estate near Paris. Despite precautions, the disease was not contained and it spread rapidly throughout France. By 1953 it had reached the Benelux countries and also Britain by unknown means; by 1954 it was in Spain, Portugal, Austria, Italy and so on. Everywhere it spectacularly reduced the rabbit population, but in Northern Europe, unlike Australia, the rabbit flea was thought mainly responsible for its transmission. The first recorded outbreak in Britain was near Edenbridge, Kent, in September 1953 but a local pest officer unfamiliar with the disease wrongly diagnosed it as syphilis. Within a month it was correctly diagnosed and an immediate attempt was made to contain it by fencing off the area, clearing vegetation, gassing warrens, etc. Despite these intensive measures the disease spread at an average rate of 3.5 miles per month in the Edenbridge area and by November 1954 there were scattered outbreaks throughout most of Britain. Some of the spread may have been deliberate as in Australia, but not in the Edenbridge area where it was due to natural causes.

There is an interesting sequel to the story. In April 1955 the first attenuated strain of the myxoma virus was found in Britain and it spread rapidly through the surviving rabbit population. The reason for its rapid spread is simple: the rabbit flea feeds only on living hosts and a rabbit infected by an attenuated strain lives longer than does one infected by a virulent strain, with a correspondingly greater chance of the attenuated strain being transmitted. There is strong evidence also that the rabbit population has changed and now includes many inherently resistant individuals as well as those that have acquired immunity. Rabbit numbers thus increased rapidly in Britain. A somewhat different situation prevailed in Australia – at least up until 1970. Attenuated strains of the virus appeared early in the 1950s, as did apparently more resistant rabbits, and yet the degree of control persisted such that in 1970 the number of rabbits was still only 1% of the 1950 level. The disease was occurring mainly as an annual mosquito-borne epizootic in the spring and summer and with a fatality rate of 30% or less. Field studies showed that this relatively low degree of control by the virus was being supplemented by the actions of predators, especially the feral cat and fox, to give the large overall degree of control. Both the highly virulent and the severely attenuated strains of the virus seem to be disappearing in Australia and they are being replaced by strains of moderate virulence. With time an equilibrium is likely to be reached in which the host resistance and virus virulence become matched. It remains to be seen at what level the rabbit population will stabilize and whether this will represent a significant reduction compared with that before the virus was introduced.

This historical interlude raises several important points. First and foremost, it shows how a pathogen, once introduced into a susceptible host population, can spread so rapidly that it cannot be contained. Second, it demonstrates the remarkable efficiency of vectors in disseminating a disease; they should be harnessed wherever possible because they are highly attuned to search out particular hosts or habitats. Third, the example shows how, with time, a host-pathogen interaction stabilizes in such a way that a proportion of the hosts are killed but a proportion remain alive. The situation in vertebrates is complicated by the immune defence system, but a similar outcome would be expected with invertebrate hosts as a result of selection pressure and because a pathogen needs a minimum host density in order to spread through a population.

Summary

Nuclear polyhedrosis viruses and granulosis viruses, which together are grouped as baculoviruses, represent the most significant prospect for insect control by micro-organisms. In both cases the virus particles are occluded within a proteinaceous shell which protects them, so they are easily stored and applied in the field. They affect mainly lepidopterous pests but also some of the Hymenoptera like sawflies, and they can cause disease of all developmental stages of their hosts. Infection occurs through the gut and leads ultimately to the release of large numbers of inclusion bodies to maintain the level of infestation of a site. The viruses have been used successfully for introductions and mass-release to give lasting control; for example against sawflies in forests and against the palm rhinoceros beetle. Also they show promise for short-term control in annual crops, and an NPV is now registered for control of the cotton bollworm in the USA. The main limitation to their further use is concern about their safety and this is exacerbated by the high frequency of inapparent infections of insects.

The myxoma virus has been introduced into rabbit populations in Australia and Europe with spectacular success, due mainly to the activities of vectors – the rabbit flea and mosquito. However, attenuated strains of the virus soon appeared, and at present it seems that a new equilibrium is evolving in which both virus virulence and the host resistance are changing in response to selection pressure and the feeding activities of the vectors.

References

Bedford, G. O. (1980). Biology, ecology, and control of palm rhinoceros beetles. *Annual Review of Entomology* 25: 309–39.

Entwistle, P. F., Adams, P. H. W. and Evans, H. F. (1977). Epizootiology of a nuclear-polyhedrosis virus in European spruce sawfly (*Gilpinia hercyniae*): the status of birds as dispersal agents of the virus during the larval season. *Journal of Invertebrate Pathology* 29: 354–60.

Fenner, F. and Ratcliffe, F. N. (1965). *Myxomatosis.* University Press, Cambridge.

Greathead, D. J. (1976). A Review of Biological Control in Western and Southern Europe. *Technical Communication 7. Commonwealth Agricultural Bureaux,* Farnham Royal.

Ignoffo, C. M. and Couch, T. L. (1981). The nucleopolyhedrosis virus of *Heliothis* species as a microbial insecticide. In *Microbial Control of Pests and Plant Diseases 1970–1980.* (Ed. H. D. Burges). Academic Press, London.

Lewis F. B. (1981). Control of the gypsy moth by a baculovirus. In *Microbial Control of Pests and Plant Diseases 1970–1980.* (Ed. H. D. Burges). Academic Press, London.

Summers, M., Engler, R., Falcon, L. A. and Vail, P. V. (Eds.) (1975). *Baculoviruses for Insect Pest Control: Safety Considerations.* American Society for Microbiology, Washington.

Tinsley, T. W. (1979). The potential of insect pathogenic viruses as pesticidal agents. *Annual Review of Entomology* 24: 63–87.

4 Microbial control of pests: use of fungi

The fungi hold a rather strange place with regard to pest control. Over 500 fungi are regularly associated with insects and some of them cause serious disease; yet very few have been used commercially as control agents and the prospects for their further use are uncertain. The main reason is that infection occurs primarily through the insect cuticle rather than through the digestive tract, so it is markedly influenced by environmental conditions. In particular a high relative humidity (at or near 100%) is always needed in the initial stages of infection and this can seldom be guaranteed in practice. Moreover, the fungal spores that initiate infections are usually very susceptible to desiccation and UV radiation, so their storage, application and persistence in the field all present special problems.

Despite the difficulties, fungi have been used successfully in a few cases and for some reason they seem to be used more frequently in the Soviet bloc than in the West. Especial attention has been given to their role in controlling aphids and scale insects, because these pests are seldom infected by other micro-organisms and some aphids have developed multiple resistance to the commonly used aphicides. The fungi are also important as natural mortality agents; *Entomophthora* spp., for example, commonly cause population crashes of aphids in the field, and this raises the question whether we can enhance the natural degree of control by appropriate manipulations.

The general features of insect-pathogenic fungi will be considered in this chapter by using as examples four fungi of particular interest. *Beauveria bassiana* and *Metarhizium anisopliae* are the two fungi most commonly used or tested in microbial control programmes; *Verticillium lecanii* has been tested extensively in the past for control of aphids and scale insects and is about to enter commercial use for aphid control in British glasshouses; *Entomophthora* spp., as noted above, have traditionally been associated with natural control of aphids in the field.

General features of insect-pathogenic fungi

Fungi infect a broader range of insects than do other micro-organisms, and infections of the Lepidoptera, Homoptera (aphids and scale insects), Hymenoptera (bees and wasps), Coleoptera (beetles) and Diptera (flies, mosquitoes) are quite common. In fact some fungi have very broad host ranges covering most of these insect groups and this is true of *B. bassiana*, *M. anisopliae* and *V. lecanii*, all of which have world-wide distributions. Some other fungi are associated with particular types of host; for example *Hirsutella thompsonii* with mites (Arachnida) and the aquatic fungi *Coelomomyces* and *Culicinomyces* spp. with mosquitoes. In yet other cases a single fungal genus contains species with both broad and narrow host ranges, and *Entomophthora* is an extreme example of this. Among its 150 or so known species, *E. sphaerosperma* has a relatively wide host range whereas *E. aphidis* and 13 other species are associated with aphids, *E. acaricida* with mites, *E. forficulae* with earwigs, and so on. It is thus difficult to generalize about the host ranges of fungi, and the reasons for differences in host specificity are unknown. In almost all cases, however, the insect-pathogenic fungi seem to be specialized for this role and they probably do not have a significant phase as saprophytes away from their insect hosts.

Several further generalizations can usefully be made as follows:

(1) Most grow readily on standard mycological media like potato-dextrose agar or malt extract agar so they are not nutritionally demanding. The exceptions include *Entomophthora* spp. which often need more complex media containing animal materials.

(2) Most grow best at 20 to 25°C and do not grow well, if at all, at 37°C. A consequence is that they present no serious hazard to man and other warm-blooded animals except in so far as some fungal spores cause allergies.

(3) They grow as typical elongate hyphae on solid substrates but in submerged culture several of them grow as yeast-like budding cells termed *blastospores*. A similar budding phase is often seen in the diseased host. Unfortunately the blastospores are usually susceptible to desiccation, so although they can be produced in large numbers in fermenters they cannot easily be used as inoculum for control programmes.

(4) The fungi readily produce asexual spores in culture or on the host in humid conditions, and in nature these are the main infective units. They are fundamentally different from bacterial endospores because they are produced in large numbers for dispersal by wind, rain-splash etc. but they have very little resistance to adverse conditions.

(5) Many fungi also produce more resistant sexual spores or other resting stages at the onset of unfavourable conditions, but the precise roles of these structures in the infection cycle are largely unknown. In many cases they probably give rise to a further batch of asexual dispersal spores under favourable conditions for disease development. It is interesting that several insect-pathogenic fungi can be isolated from soil or organic debris on to selective media (e.g. Doberski & Tribe, 1980) and it seems likely that their presence and survival in these environments are a result of resting bodies.

Taxonomy Insect-pathogenic fungi are placed primarily, though not exclusively, in the *Deuteromycotina* or 'imperfect fungi'. This is a rather loose grouping used for fungi that seldom if ever produce sexual stages and it includes the more familiar examples *Penicillium* and *Aspergillus*. The genera within it are distinguished by morphological and developmental features of the spore-bearing structures (the spore-bearing hyphae being termed *conidiophores*) and the species are distinguished by differences in for example size and shape of the spores (*conidia*). In *B. bassiana* (Figure 7) the conidia arise as blown-out ends of cells that grow sympodially and are swollen at the base (the generic features); the conidia themselves are small (about 3 μm), colourless and globose. In *M. anisopliae* (Figure 7) the conidia arise from special spore-producing cells termed phialides that are borne on much-branched conidiophores; the spores in this case are elongated, 10 to 14 × 3 to 4 μm, and often green. In *V. lecanii* (Figure 7) the spores are again produced from phialides but these arise in whorls from the conidiophore; the conidia are colourless and range in size from 3 × 1.5 up to 10 × 2.5 μm. Incidentally, the range of spore size in this case has been used by some people as a basis for distinguishing several different species in what is here termed '*V. lecanii*'. We need not concern ourselves with the arguments for and against this except to note that separation would be necessary if spore size were found to be related to differences in biology.

The genus *Entomophthora* is quite different from those above. It is classified in the *Zygomycotina*, a grouping that includes *Mucor* and *Rhizopus* and in which the asexual spores characteristically are formed by cytoplasmic cleavage within a *sporangium*. However, *Entomophthora* differs from the norm in this respect because cytoplasmic cleavage does not occur and instead the whole sporangium is shot off from the supporting hypha and functions as a spore.

Further details of the taxonomy of insect-pathogenic fungi can be found in the review by Ferron (1978).

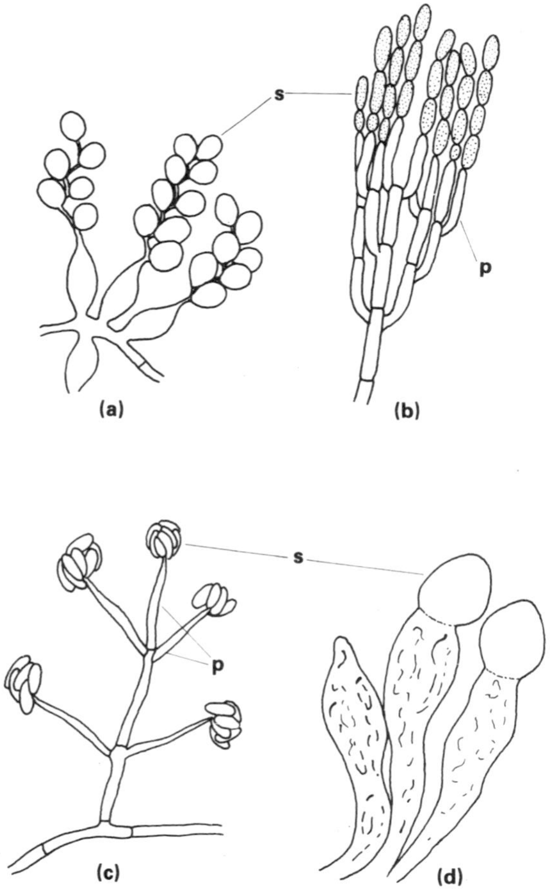

Fig. 7 Insect-pathogenic fungi. **a** *Beauveria bassiana*; **b** *Metarhizium anisopliae*; **c** *Verticillium lecanii*; **d** *Entomophthora* sp. Spores (s) and phialides (p) are indicated.

Infection and disease development In suitably humid conditions a spore present on the insect cuticle germinates to produce a young hypha, termed a *germ tube*, and this sometimes swells at its tip to form an anchoring structure, the *appressorium*. (In *Entomophthora* the appressorium is formed directly from the spore.) A thin *penetration hypha* develops beneath this; it can either penetrate both the epicuticle and procuticle or it penetrates only the epicuticle, then expands into normal-sized hyphae that run longitudinally and give rise to further penetration hyphae. A high humidity is needed only for these very early stages of infection. If the insect moults at this stage the infection can be lost; otherwise the fungus continues to grow and invades the epidermis and hypodermis. This is the most common stage at which the host defence mechanisms are seen to operate, and local tissue reactions can contain the pathogen at the initial sites of infection. If the fungus overcomes the defences it frequently gives rise to blastospores or hyphal fragments (so-called *hyphal bodies*) which circulate and proliferate in the haemolymph. The insect then dies quite rapidly. It is uncommon for the major organs to be colonized before death (the fat body is the first organ to be colonized in such cases) but after the host dies the fungus commonly reverts to a typical hyphal form and extensively colonizes the major organs. This is often termed the 'saprophytic' phase of development and it is followed, in humid conditions, by the outgrowth of some of the hyphae through the cuticle. These hyphae produce numerous dispersal spores and may completely cover the surface of the cadaver. Alternatively the fungus can form resting stages within the dead host in conditions unsuitable for further spread of an epizootic, and the resting stages probably serve in over-wintering. The whole sequence is summarized below, where brackets indicate the stages that are not always found.

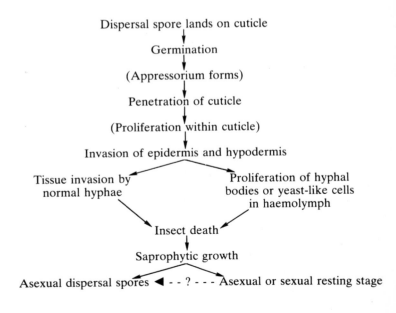

Dispersal spore lands on cuticle
↓
Germination
↓
(Appressorium forms)
↓
Penetration of cuticle
↓
(Proliferation within cuticle)
↓
Invasion of epidermis and hypodermis
↙ ↘
Tissue invasion by normal hyphae | Proliferation of hyphal bodies or yeast-like cells in haemolymph
↘ ↙
Insect death
↓
Saprophytic growth
↙ ↘
Asexual dispersal spores ◀ - - ? - - - Asexual or sexual resting stage

Several features of this infection process merit further comment. The phase of penetration involving an appressorium and a thin penetration hypha is identical to that in plant pathogenesis, in which for economic reasons it has received much closer study. The formation of appressoria on plants has been related to the presence of both stimulatory and inhibitory materials that can be extracted from the cuticle and surface waxes, but there is little if any evidence of specificity at this stage of infection because appressoria can be formed on hosts and non-hosts alike. It is interesting that inhibitory and stimulatory materials have also been extracted from insect cuticle. The anchoring material beneath the appressorium has been identified as a hemicellulose in the case of one plant pathogen, but no equivalent identification seems to have been made for insect pathogens. The penetration process in plants is thought to be brought about by a combination of mechanical forces at the hyphal apex and enzymic degradation of the plant cell wall. The same is probably true in insect penetration and in this respect it is notable that the pathogens produce lipases and proteases in culture, and most (not *Entomophthora*) also produce chitinase. This last enzyme is perhaps the least important in penetration because chitin does not confer structural rigidity on the cuticle; this is achieved by cross-linkages between the protein components, cleavage of which would not require a chitinase.

Perhaps the most intriguing and problematical aspect of insect pathogenesis concerns the role of toxins. In culture both *B. bassiana* and *M. anisopliae* produce depsipeptide toxins that are active on injection into insects; the structures of two examples, *beauvericin* and *destruxin B*, are shown below. The destruxins are thought to play a significant role in disease because the host dies early in the course of infection, before the fungus has made substantial growth within the tissues. The role of beauvericin is less clear because *B. bassiana* makes more substantial growth by the time that the host dies and pathogenicity of strains is not always closely correlated with their abilities to produce the toxin *in vitro*. *V. lecanii* also is reported to produce a toxin – the depsipeptide *bassianolide*, so-called because it was first reported from *B. bassiana*. In contrast, *Entomophthora* spp. probably do not produce toxins (those that have been reported are proteins and probably represent enzymes involved in tissue degradation) and this is consistent with the fact that these fungi usually grow very extensively in the tissues while the host is still alive.

Some of the toxins like beauvericin have weak anti-bacterial activity and this has led to the suggestion that they serve to restrict bacterial growth in the dead or dying insect so that the fungus can grow unimpeded in its saprophytic phase. Clearly, more work is needed on the role of toxins, not least because it could be important for strain selection.

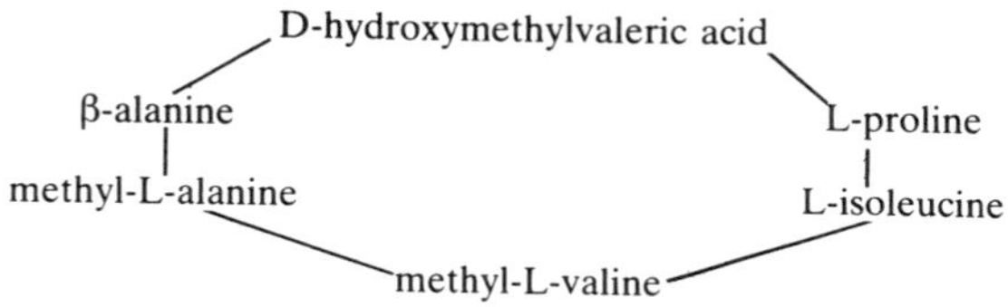

Destruxin B *(Metarhizium anisopliae)*

L-N-methylphenylalanine

D-2-hydroxyisovaleric acid — D-2-hydroxyisovaleric acid

L-N-methyl phenylalanine — L-N-methyl phenylalanine

D-2-hydroxyisovaleric acid

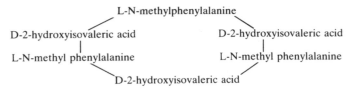

Beauvericin *(Beauveria bassiana)*

Applications for microbial control

Most attempts to use fungi have involved adding relatively large amounts of inoculum to achieve short-term, season-long control. The following examples are typical of this approach.

B. bassiana is reported to be used on over 70 types of crop in the Soviet Union, but the details are difficult to obtain and the effectiveness of the control measure is hard to assess. The inoculum is prepared by growing the fungus as blastospores in aerated submerged culture and then spreading them on to sterilized bran or cereal grains in trays. The solid substrates induce normal hyphal growth and subsequent sporulation. The conidia are then harvested, dried, mixed with kaolin as an inert carrier and distributed as a powder termed *Boverin*. A good example of its use is for control of the Colorado beetle *Leptinotarsa decemlineata* in potato crops. According to Ferron (1978), Boverin is applied at a rate of 2 kg/hectare (spore concentration 6×10^9/g) together with one fifth of the normal dose of DDT or other insecticide, in two treatments spaced 15 to 21 days apart. Alternatively the fungus is used alone at 3 to 4 times the above rate, and this now seems to be the preferred method. The aim of the combined treatment is to stress the insect with a sub-lethal dose of insecticide so that it is more susceptible to the pathogen. There is evidence for this type of effect in some cases but recent work suggests that *B. bassiana* and DDT have separate and complementary effects rather than acting synergistically.

M. anisopliae has been tested very extensively in the past against a range of pests but its best known use at present is for control of the froghopper *Mahanarva posticata* in sugar cane plantations in Brazil. For this the fungus is grown on boiled rice grains until it sporulates; the rice is then dried to 35% RH and ground to a powder. The resulting product, termed *Metaquino*, can be stored at 7°C without loss of viability or pathogenicity. The fungus also has been used to control the rhinoceros beetle in Tonga by applying spores to sawdust heaps in which the beetle larvae develop, but in general the baculovirus seems to have given better control of this pest (see above, pp. 24–5).

V. lecanii has been used experimentally and on a small commercial scale to control aphids and scale insects on a range of plants throughout the world; in most of the literature on this it is called *Cephalosporium lecanii* (e.g. Baird, 1958). Its use continues to the present day – for example to control the coffee green bug *Coccus viridis* in India; but parasitic insects have in general been more successful for control of scales, and most recent attention has therefore focused on the potential of *V. lecanii* for aphid control. It is about to enter into commercial use for control of aphids on the year-round chrysanthemum crop in British glasshouses. A commercial formulation for this purpose, named *Vertalec*, was developed by Tate & Lyle in Britain and is marketed by Koppert BV, a Dutch-based company.

The year-round chrysanthemum is one of the most important British glasshouse crops and an ideal one on which to use fungal control agents. It is grown at a minimum of 15°C which is near the optimum temperature for growth of *V. lecanii* (about 20°C). More importantly, the crop is 'blacked out' with polythene sheeting from mid-afternoon until morning during the summer to control the initiation of flowering. This creates a high enough humidity for fungal spores to germinate and infect aphids. Hall & Burges (1979) showed that a single spray of conidia applied just before black-out gave satisfactory control of the important aphid *Myzus persicae* but less so of two minor aphid pests *Brachycaudus helichrysi* and *Macrosiphoniella sanborni* in small experimental glasshouses. Control was achieved within 2 to 3 weeks of spraying and it lasted for the duration of the crop. The difference in degree of control of the three aphids is interesting because it could not be explained by differences in susceptibility of the pests; in fact *M. persicae* was slightly less susceptible than the others in laboratory tests. The proposed explanation centres around ecological and behavioural differences between the aphids. *M. persicae* feeds mainly on the undersides of leaves where the humidity is likely to remain highest and it also tends to be more mobile than the others on chrysanthemums which are a relatively unfavourable food crop for it, so it probably spreads the infection by contagion.

V. lecanii is interesting for a quite different reason because it is reported as a parasite of other fungi. Specifically, it grows on the pustules formed by some plant parasites as they break through the plant epidermis to sporulate – for example the dwarf bean rust fungus *Uromyces appendiculatus*, the carnation rust fungus *Uromyces dianthi* and the wheat stem rust fungus *Puccinia graminis*. When applied to carnation leaves in experimental conditions it reduced the number of pustules formed by *U. dianthi*. So there is the interesting possibility that one fungus might be used to control both the pests and pathogens of a crop, though work on this is still in its infancy. Actually it is unusual for a fungus to parasitize both insects and other fungi, but it is not entirely unexpected. Fungal walls and insect cuticle both contain chitin (though not necessarily as the main component) and fungi and insects have the same storage compounds (glycogen, lipids) and soluble carbohydrates (mannitol and trehalose), so as hosts of *V. lecanii* they have several features in common. In a study of 8 isolates of *V. lecanii* from fungal hosts, 3 were highly pathogenic to aphids, 3 were weakly and atypically pathogenic (a low proportion of the aphids died and these often had secondary bacterial invaders) and 2 were non-pathogenic to aphids. So there seems to be no clear distinction between isolates of *V. lecanii* from fungal and insect hosts, just a large degree of variation which is quite common among the insect-pathogenic fungi in general.

Fungi as natural mortality agents

Aphids in the field are frequently infected by *Entomophthora* spp. and in some years this may result in a significant degree of control. The factors that influence these epizootics are very complicated; they include the availability of inoculum early in the season, the aphid population density, and climatic factors which influence both the rate of multiplication of the host and the activity of the fungus.

In fact epizootics frequently develop late in the growing season when the pest has already attained damaging levels, as is shown in Figure 8. It is not simply a result of the slow progress of disease in individual aphids, because death can occur within 3 to 5 days at 20°C and within 10 to 11 days at 10°C.

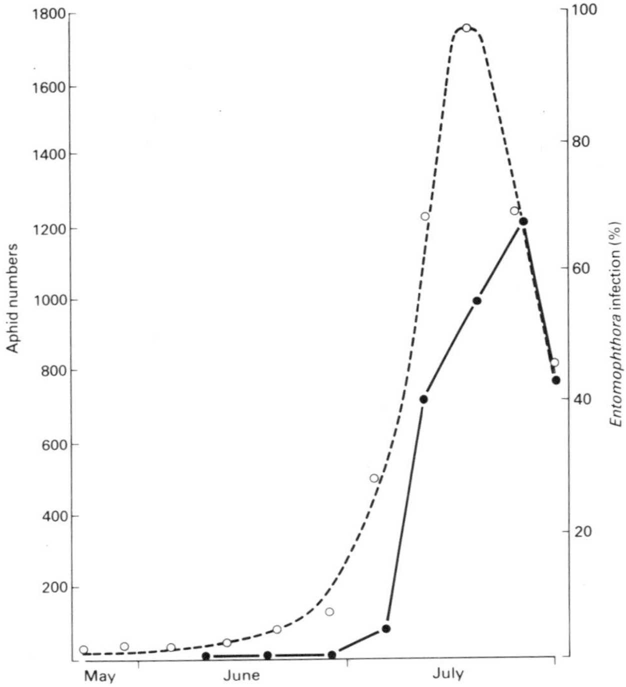

Fig. 8 Numbers of aphids (*Aphis fabae*) in a bean crop during the course of a season in Britain (broken line) and percent infection by *Entomophthora* (solid line). Note that infection by *Entomophthora* rises to substantial levels only after the aphid population has attained a high level. (Based on results in Wilding & Perry (1980) by kind permission of the authors and the publisher.)

Various attempts have been made to enhance the natural degree of control by applying spores of *Entomophthora* early in the season, but this is impracticable even if the inoculum level is the critical factor limiting the development of an epizootic. In a study of three *Entomophthora* spp. the spores germinated only at or above 95% RH and they lost viability if kept for more than 6 h at 80% RH. For the same three species germination was negligible at 8°C, so both the temperature and the humidity must be predictable for inoculation to have any chance of success. An alternative is to use resting spores (*chlamydospores*) as an inoculum source; unlike the dispersal spores these are directly equivalent to bacterial endospores in both development and function and they are probably the main means of over-wintering in field conditions. The possibility of their use has been brought one stage nearer by the production of relatively large numbers of resting spores in inexpensive media in fermenters; for example 3×10^6 spores/ml in the case of *E. virulenta*. But this is still not possible for all species and there is a further difficulty that the spores so formed do not germinate readily. A particularly interesting feature of these resting spores is that their physiology seems to be highly attuned to the life cycle of the host. For example Wallace *et al.* (1976) found that a photoperiod of 14 h is necessary to trigger germination, and this would have the effect that the spore does not germinate in conditions in which the crop plant and thus the aphid are unlikely to be active. Species and strains from different geographical regions are likely to differ in this respect, so there is a possibility of introducing strains that will germinate earlier than do the resident ones in a field site.

We saw earlier in the case of *V. lecanii* that a pathogen may be present and yet give unsatisfactory control because of behavioural or ecological features of the insect host. An example of this in natural environments is provided by the elm bark beetles *Scolytus scolytus* and *S. multistriatus* which are of interest because they are vectors of Dutch elm disease. Doberski & Tribe (1980) found diseased beetles only occasionally in a survey of dead trees in the south of England, and yet the fungus *B. bassiana* was isolated from 81% of all bark samples and from 94% of soil samples taken at the tree bases. (Incidentally *M. anisopliae* was isolated only once in this study.) Spores of *B. bassiana* infect both adult beetles and larvae in laboratory conditions, so the low natural level of infection must be explained on other grounds. A likely explanation is that the beetle galleries in the bark (the so-called brood galleries) seldom interconnect: as shown in Figure 9, each larva eats out a channel within the bark and these diverge from one another. The chances of cross-infection are thus minimal and the only real prospect of controlling the pest with a microbial agent is if this can be introduced with the adult female as it lays the eggs so that all the emerging larvae are contaminated. Webber (1981) has shown that elm bark colonized by the fungus *Phomopsis oblonga* is inimical to the breeding of *Scolytus* beetles, and it is suggested that this represents a natural biological control of the vector of Dutch elm disease. Interesting though it is, the evidence is inconclusive as yet because it shows merely a correlation between presence of the fungus and absence of beetle breeding; until a causal relationship is demonstrated there remains the possibility (or even likelihood) that some other factor like water content of bark independently affects both the fungus and the beetle.

Attention has been given to the possibility that man might inadvertently destroy the natural degree of microbial control by fungi in his use of chemicals. Returning to the experiments on three *Entomophthora* spp. mentioned earlier, the insecticides Methoxychlor, Malathion and Diazinon were found to have no adverse effects on spore germination whereas five fungicides (Captan, Dodine, Ferbam, Maneb and Zineb) all significantly reduced germination. The significance of this is that agricultural advisors usually specialize in either pest or disease control so they might be unaware of the consequences of their advice for the other discipline.

Advantages and disadvantages in the use of fungi

Advantages (1) Many insects are parasitized by one fungus or another and in some cases fungi are the only potential microbial control agents. (2) The fungi

Fig. 9 Breeding galleries of *Scolytus multistriatus* in elm bark. The adult female beetle eats out a large central channel and lays eggs in it; the emerging larvae eat out radiating channels which often end in oval pupal chambers. (Taken from Webber (1981) by kind permission of the author and the publisher.)

generally infect all developmental stages of their hosts so they can be applied at any appropriate stage. (3) Some fungal pathogens have very broad host ranges and should prove especially attractive as control agents if the problems of inoculum production and storage can be overcome. This is already seen in so far as *B. bassiana* is used very widely in the Soviet Union to control a range of pests. (4) Fungi generally present no health hazard to man and other vertebrates and there are no reports that they harm the natural enemy complex (though equally there is no reason why they should not do so). (5) They cause rapid death of their hosts, followed by abundant sporulation, so in appropriate conditions they can cause very damaging epizootics. (6) They are generally compatible with insecticides and in some cases act synergistically with them.

Disadvantages (1) The most serious disadvantage is the difficulty of obtaining satisfactory types of inoculum for use in field conditions, because of the sensitivity of fungal spores to desiccation and UV radiation. Nevertheless, the fungi evidently persist quite well in soil and organic debris so further study of the mechanisms of survival may lead to different approaches for introductions. (2) Epizootics caused by fungi are influenced by a complex of environmental factors, so it is difficult to predict the success of inoculations. (3) Many insect-pathogenic fungi are susceptible to the commonly used fungicides for plant disease control. This could possibly be overcome by mutation and strain selection in the laboratory because spontaneous fungicide-resistance is quite common (amongst plant pathogens) and is normally a result of failure to take up the fungicide.

Outstanding problems and future research

Undoubtedly the main problems concern inoculum production, storage and application in the field. There is scope for developing types of inoculum based on resting spores rather than dispersal spores in some cases (e.g. *Entomophthora* and *M. anisopliae*, but not *B. bassiana* which apparently does not produce them) and the potential success of these will depend on a better understanding than at present of their behaviour and functions in relation to disease. For example, can they germinate on the insect cuticle or in the gut and cause infection directly? The evidence for *Entomophthora* suggests that they do not, but rather that they germinate usually to produce one or more dispersal spores, and even then only after a period of constitutive dormancy.

In some cases it may be better to seek alternative ways of introducing fungi into the host population, for example by releasing infected adults or by using attractants to lure insects towards an inoculum source. A different approach is to recognize that pathogenic fungi are already very widely distributed but their activities are limited by environmental factors. Thus it may be more fruitful to seek ways of altering the microclimate in which the pest exists so that introduced or natural inoculum is more effective. This can be done relatively easily and cheaply in glasshouses, though we should not lose sight of the fact that pest control is only one of several factors that a grower must take into account (see below, pp. 82–3). It is also quite feasible in the field, for example by altering plant density or growth form or by changes in husbandry practices like the timing and method of irrigation where it is used.

Summary

A few fungi are used commercially for pest control, the main ones being *Beauveria bassiana, Metarhizium anisopliae* and *Verticillium lecanii*. These have very wide host ranges covering several insect groups and they infect at all stages of the host's development. In addition, fungi are important natural mortality agents – for example *Entomophthora* spp. on aphids – but epizootics frequently occur late in the season after the pest has caused economic damage. Fungi normally infect through the insect cuticle and cause rapid death, sometimes probably as a result of toxins. The spores that initiate these infections are produced in large numbers on the dead or dying host and are often aerially dispersed. However, they are susceptible to desiccation and this seriously limits their use as inoculum for control programmes. A high relative humidity is needed in the initial stages of infection, and environmental conditions in general limit the effectiveness of fungal control agents. Nevertheless, fungi have great potential for pest control if these specific problems can be overcome, and for some pests they are the only potential microbial control agents.

References

BAIRD, R. B. (1958). *The Artificial Control of Insects by Means of Entomogenous Fungi: a Compilation of References with Abstracts*. Entomology Laboratory, Belleville, Ontario.

DOBERSKI, J. W. and TRIBE, H. T. (1980). Isolation of entomogenous fungi from elm bark and soil with reference to ecology of *Beauveria bassiana* and *Metarhizium anisopliae*. *Transactions of the British Mycological Society* 74: 95–100.

FERRON, P. (1978). Biological control of insect pests by entomogenous fungi. *Annual Review of Entomology* 23: 409–42.

HALL, R. A. and BURGES, H. D. (1979). Control of aphids in glasshouses with the fungus *Verticillium lecanii*. *Annals of Applied Biology* 93: 235–46.

WALLACE, D. R., MACLEOD, D. M., SULLIVAN, C. R., TYRRELL, D. and DELYZER, A. J. (1976). Induction of resting spore germination in *Entomophthora aphidis* by long-day light conditions. *Canadian Journal of Botany* 54: 1410–18.

WEBBER, J. (1981). A natural biological control of Dutch elm disease. *Nature* 292: 449–51.

WILDING, N. and PERRY, J. N. (1980). Studies on *Entomophthora* in populations of *Aphis fabae* on field beans. *Annals of Applied Biology* 94: 367–78.

5 Disease control: use of specific microbial agents

Plant diseases are caused primarily by micro-organisms, though microbiologists seem to have recognized this only recently. Fungi and viruses are the most important plant pathogens, with bacteria coming a poor third for most crops (Table 4). Examples of microbial control against all these groups of microbes will be considered, but inevitably most attention will be focused on fungal pathogens because of their importance and their suitability for microbial control.

Perhaps the most serious problem in microbial control of diseases is the fact that pathogens characteristically grow within the tissues of their hosts and thus in highly selective environments from which most potential control agents are excluded. Control must therefore be attempted either in the early stages of infection while the pathogen is on the host surface, or in the pathogen's dormant or saprophytic phase in soil or crop residues. This in turn leads us into the realm of microbial interactions in natural communities – an exceedingly complex area of which our understanding is far from complete. In short, we shall be dealing with a quite different situation from that in pest control because we shall need to consider not only the pathogen and the control agent but also the other members of the microbial community and their effects on the interaction.

In the first of these chapters we consider relatively simple cases – those in which a specific micro-organism is introduced to achieve disease control. Subsequent chapters consider cases in which the environment is altered in some way to promote the activities of existing control agents and finally those in which naturally occurring microbial control is being exploited with a minimum of interference from man.

Definition of terms

The following are working definitions for use in the chapters on disease control and they should not be regarded as absolute.

Antagonism The *direct* detrimental effect of one organism (*antagonist*) on another; for example by *antibiosis* (antibiotic-production), *parasitism* (feeding) or *predation* (capture).

Competition The *indirect* detrimental effect of one organism (*competitor*) on another, for example by using its food source or infection site, etc.

Lysis The partial or complete *enzymic* breakdown of an organism. In some cases the enzymes are produced by another organism (an antagonist) in which case the phenomenon is termed *heterolysis* or *exolysis*. In other cases the enzymes are produced by the affected organism (*autolysis* or *endolysis*), usually in response to nutrient-stress.

Parasitism A nutritional relationship in which one organism (*parasite*) gets all or part of its needs from the living functioning parts of another organism (*host*). If the parasite causes disease it is termed a *pathogen*, so we can get pathogenic and non-pathogenic parasites. Pathogens are normally *virulent*, but *avirulent* or *hypovirulent* strains or mutants can occur. Thus we can regard pathogenicity as

Table 4 Estimated contributions of fungi, bacteria and viruses to losses from crop disease in the USA

Crop	% Total disease loss caused by				Main pathogens *
	Fungi	Bacteria	Viruses	Unspecified †	
Wheat	70.6	0	7.1	22.3	*Puccinia graminis* (f)
Oats	47.5	0	18.3	34.4	Yellow dwarf (v), *Puccinia coronata* (f)
Barley	44.3	0	34.1	21.4	Stripe mosaic (v), *Helminthosporium* (f)
Rice	62.9	0	0	37.1	*Piricularia oryzae* (f)
Maize	51.7	4.2	0	44.1	*Diplodia maydis* (f), *Gibberella zeae* (f)
Potatoes	48.5	13.1	31.6	6.8	*Phytophthora infestans* (f), leaf roll (v)
Sugar	25.0	0	(56.3)	18.7	Yellows (mycoplasma), *Cercospora beticola* (f)
Tomatoes	43.0	4.8	28.6	23.6	Tobacco mosaic (v), *Stemphylium* (f), *Verticillium* (f)
Apples	71.0	15.0	2.0	12.0	*Venturia inaequalis* (f), *Erwinia amylovora* (b)
Peaches	56.0	25.0	13.0	6.0	*Monilinia fructigena* (f), *Xanthomonas pruni* (b)
Tobacco	48.2	21.8	12.7	37.3	Tobacco mosaic (v), *Pseudomonas tabaci* (b)
Average	51.7	7.6	18.5		

* Fungus, f; bacterium, b; virus, v.
† Includes disease complexes like 'root rot' caused mainly by fungi.

(Based on United States Department of Agriculture (1965). Losses in agriculture. *Agricultural Research Service, Agriculture Handbook Number* **291,** Washington.)

a feature of an organism like a species, but within a species the strains may differ in virulence.

Use of *Peniophora gigantea* to control *Heterobasidion annosum*

The fungus *Heterobasidion annosum* (formerly *Fomes annosus*) is a major pathogen of coniferous trees in the Northern Hemisphere, and the most serious pathogen in British forests. It is a root-infecting fungus that spreads relatively slowly along the tree roots by hyphal growth and can cross from one tree to another by root-to-root contact. In larch and spruce it causes serious economic losses as a result of *butt rot*, when it grows from the roots into the tree base and decays the non-living heartwood as a saprophyte. In pines, however, it seldom causes a butt rot; instead it is most damaging as a parasite because it kills the roots and eventually kills the tree itself at any stage of tree development.

H. annosum produces bracket-shaped fruiting bodies at the base of infected trees, and they release *basidiospores* which are wind-dispersed. The spores play only a minor role in the infection cycle in normal conditions because they are small (4.5×3.5 μm) and have very low nutrient reserves, so they are unable to invade through the outer corky tissues of the exposed parts of roots. This situation is altered drastically in commercial forestry when trees are felled for harvest or when plantations are thinned. Fresh living cells are then exposed on the resulting stump surfaces and the stumps as a whole can remain alive but with reduced host resistance for several months, creating a selective environment for pathogens or other parasites. *H. annosum* colonizes the stump surface from basidiospores in the air and grows down into the attached roots and then to the roots of adjacent living trees. It is virtually impossible to eradicate once established in a site, so control measures are aimed at protecting the stumps from invasion. This is particularly important in Britain because most of the forests have been established in the last few decades on previously tree-less sites where there was no resident inoculum of *H. annosum* in the soil.

Various stump treatments have been tried. Early attempts involved sealing the stump surface with creosote or other tar-based materials and this generally worked well until the stumps began to crack and dry, exposing fresh surfaces. Phytotoxic chemicals can be used and sometimes are the most effective treatment. They kill the stump surface tissues, hastening colonization by saprophytes which can then exclude *H. annosum*. Indeed they are necessary for broad-leaved trees to prevent re-growth from the stumps – a problem that does not occur with most conifers. They are mostly fairly innocuous compounds like nitrites, urea and boron-containing materials; but they have the disadvantage that they affect only the surface tissues, so if *H. annosum* is already present on the roots it can spread to adjacent trees and can also grow up to the stump surface and sporulate, maintaining the inoculum level in the air.

Rishbeth (1963) pioneered a new method of stump protection for pines, following his observation that another fungus *Peniophora gigantea* sometimes colonizes pine stumps naturally in competition with *H. annosum* and excludes it. Indeed part of the success of ammonium sulphamate treatment of stumps as a chemical control measure could be attributed to its effect in promoting colonization by *Peniophora*. The problem was that the level of natural colonization was both too low and too erratic to be generally effective, but this could be overcome by inoculation. Now the inoculation procedure is used routinely on pines in Britain – on a total of 62 000 hectares in 1973.

Production and use of inoculum *P. gigantea* is somewhat unusual amongst members of the fungal group *Basidiomycotina* (the higher fungi, including the toadstool-producers) because it forms abundant asexual spores in culture. These are brick-shaped and are formed by fragmentation of the hyphae; they are loosely termed *oidia* or *arthrospores* though they are, in fact, conidia. For early experimental control programmes they were harvested from cultures and constituted into tablets which could be dispersed in water for spraying on to the stumps. A liquid spore inoculum is preferable but in this case it was difficult to obtain because the spores germinate readily in water. The problem was overcome in a simple and elegant way by adding sucrose to lower the water activity (osmotic potential) below the point at which germination occurs. Now the fungus is available in sachets, each containing not less than 5×10^6 viable spores suspended in 1 ml sucrose solution with a dye. The sachets can be stored for up to 6 months in a refrigerator. For use, the contents are added to 5 litres of water and the diluted spore suspension is applied to freshly cut stump surfaces from a plastic bottle with a brush inserted in the lid. Each sachet contains enough inoculum to treat about 100 stumps, at a cost of a few pence per stump including labour.

The results of this inoculation programme have been spectacular. The degree of stump protection at least equals that achieved with chemicals, and one year after treatment the whole stump surface is often covered with sporing structures of *Peniophora*. The spore load in the air is sufficiently increased, while that of *H. annosum* is decreased, for the fungus to be thought to bring about an enhanced degree of natural control as well as that achieved by inoculation. The rationale behind the method is to add an overwhelming inoculum of *Peniophora* so that this fungus colonizes the stump surface rapidly and completely. Being a parasite, *Peniophora* can overcome the residual host resistance in the stump tissues, but it is not sufficiently aggressive to be a pathogen and so it poses no threat to neighbouring healthy trees. The treatment has two significant advantages over chemical stump treatment. First, *Peniophora* can grow into any parts of the stump missed in the initial application. Second, it can grow down into the stump tissues and localize any resident colonies of *H. annosum*. Its role in disease control is summarized in Figure 10. However, the treatment of individual stumps is inconvenient, so current research in Europe and the USA is focused on the possibility of adding *Peniophora* spores to the oil used to lubricate chainsaw blades, so that the processes of felling and inoculation can be combined. This seems to work well if a light oil (SAE 10) is used but not with SAE 30.

Unfortunately, *Peniophora* can be used only on pine stumps and not on those of, for example, sitka spruce which is currently the most commonly planted tree in Britain. Further research is under way to try to find equivalent fungi for use on other trees, but as yet none has reached the stage of commercial exploitation.

Mode of action Ironically the mode of action of *Peniophora* came to light after the fungus had been used successfully as a control agent, and in a totally unrelated study – the succession of fungi on dung of herbivores. It is well known (truly!) that

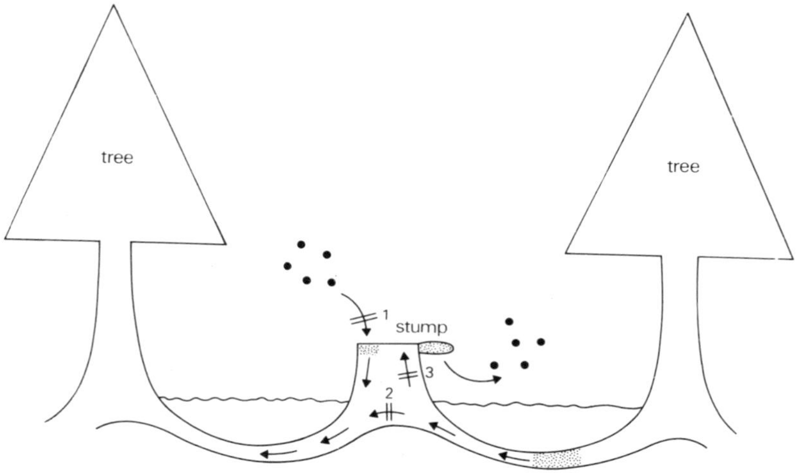

Fig. 10 Mode of action of *Peniophora gigantea* against *Heterobasidion annosum* (stippled) in pine stumps. (1) *Peniophora* prevents the pathogen from colonizing the stump from air-borne spores and then travelling down into the root zone. (2) and (3) When established in the stump *Peniophora* can restrict spread of the pathogen from existing foci of infection, so that it cannot colonize other roots (2) or grow up to the stump surface and sporulate (3).

members of the Basidiomycotina are the last to appear when dung is incubated in the laboratory, and when they appear they usually suppress the fruiting of other fungi. This was explained by Ikediugwu & Webster (1970) who showed that many of the Basidiomycotina antagonize other fungi on contact or at very close range. The phenomenon is termed *hyphal interference*. When the basidiomycotina are paired with each other on agar plates they are seen to have different degrees of interfering activity, and in this respect *Peniophora* is seen to interfere strongly with *Heterobasidion*, paralleling what is thought to occur in stump tissue. The phenomenon itself is most interesting. A hypha of *Heterobasidion* stops growing on making contact with *Peniophora*; the cytoplasm in the tip appears granular at first and then highly vacuolated, and the hypha takes up dyes like neutral red which previously were excluded from it (Figure 11). This suggests a change in membrane integrity. The mitochondria in the affected cells become swollen and rounded and subsequently degenerate, the cytoplasm develops dense barrier regions (apparently to localize the damage) and a wide 'extraplasmalemmal zone' develops between the plasmalemma and the cell wall (Ikediugwu, 1976). As indicated above, these events are strictly localized in the affected hypha; often they occur only in the cell (actually a compartment) in contact with the interfering hypha. There is no evidence that *Peniophora* or other interfering species actually parasitize other fungi. Rather, hyphal interference seems to be a means of killing or immobilizing hyphae of other fungi that are potential competitors for the same substrates. The underlying mechanism is almost completely unknown and must surely be one of the most fruitful topics for study.

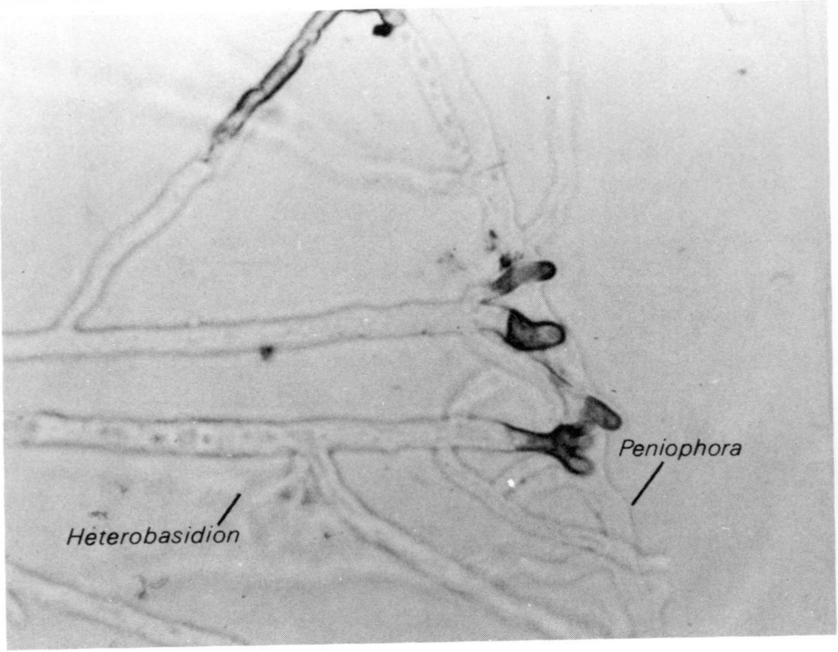

Fig. 11 Interaction between *Peniophora gigantea* and *Heterobasidion annosum* on agar. The agar plate was flooded with neutral red; the dye has been taken up by hyphae of *Heterobasidion* where they contact a hypha of *Peniophora*.

Control of crown gall by *Agrobacterium radiobacter*

Crown gall is a disease of many woody plants like cherry, peach, apple, roses, hops and vines. It is caused by the bacterium *Agrobacterium tumefaciens* which enters through wounds and it is characterized by the production of tumour-like swellings (galls) on the roots or stems (Figure 12) and especially at the *crown* of the plant, just above soil level. It causes economic losses mainly because the plants are unsaleable or because their movement is restricted by quarantine regulations. A remarkable feature of the disease is that plants can bear several galls but only the first-formed one need contain bacteria. The others are often microbiologically sterile and their cells can be cultured in the laboratory without the need to supply plant hormones, unlike most normal plant cells. This led to the early recognition that a *tumour-inducing principle* is involved, and recently this has been identified as a piece of bacterial DNA.

Both tumorigenic and non-tumorigenic strains of the pathogen *A. tumefaciens* occur, the difference being that only the tumorigenic ones contain a plasmid termed the *Ti plasmid* (tumour-inducing) of 112 to 156 × 10^6 molecular weight. A plasmid is a circular DNA molecule that normally exists independently of the chromosome, and some plasmids like Ti are termed *conjugative* plasmids because they code for their ability to be transferred from one bacterial cell to another during mating (Hardy, 1981). Thus non-tumorigenic strains of *A. tumefaciens* become tumorigenic when they acquire the plasmid and vice versa. In 1977 it was shown beyond doubt that plant cells also can receive part of the plasmid and they then give rise to galls. The cells in a gall are estimated to contain about 20 copies of a plasmid segment of 3.7 to 6.0 × 10^6 molecular weight, i.e. enough to code for about 10 proteins. This part of the plasmid clearly contains the genes coding for

Fig. 12 Crown gall, caused by *Agrobacterium tumefaciens*, on stem of blackberry (*Rubus* sp.).

tumour production. It also contains the gene coding for the production of opines, like *nopaline*, which are unique amino acid derivatives. This is interesting because opines are produced only in gall tissue and cannot be used by the plant itself; but they can be used by the bacterium as sole energy sources. In other words the pathogen 'programmes' the plant to produce a specific nutrient source.

The Ti plasmid codes for two other features of special relevance. The first is the ability of the bacterium to adhere to infection sites (wounds), and the second, the sensitivity of the bacterium to a *bacteriocin* (an antibiotic) produced by one particular strain of the closely related but non-pathogenic bacterium *Agrobacterium radiobacter*. This organism is used as a microbial control agent.

Development of a microbial control programme Kerr and his colleagues in Australia pioneered the present approach to the control of crown gall in 1972. They isolated a strain of *A. radiobacter* termed K84 that could be applied to peach seeds to protect them from crown gall when planted into pathogen-infested soil. Three months after planting, 31% of treated plants had crown gall symptoms compared with 79% of untreated ones. Similar results using this strain have been obtained in many parts of the world on a wide range of plant species, and strain 84 is now used as a commercial control agent in several countries (Moore, 1979). For practical use the bacterium is grown on agar or in liquid culture and the cells are harvested and suspended in a finely ground peat base which can be applied with the seeds at sowing. This type of inoculum is exactly the same as the *Rhizobium* inoculum marketed for nodulation of legumes. Alternatively the bacterium can be used as a suspension in water, with or without a 'sticker' like carboxymethylcellulose (the basic component of wallpaper paste). In this form it is applied as a spray or as a root-dip treatment for cuttings or young rooted plants when they are transferred from the nursery bed to their final planting sites. Growers can prepare the suspensions themselves from agar plates which are supplied by research institutes and can be stored for several weeks in a refrigerator without serious loss of viability.

Control in practice has been highly successful, whereas previously there was no satisfactory control measure. For example the incidence of crown gall on roses in Australia has been reduced from damaging levels to less than 1% and some rose growers have returned to land previously abandoned for rose-growing because of the disease. Similar success has been reported in parts of Europe and the USA. However, not all results have been so spectacular, the reason being that only some forms of the pathogen are subject to control. We shall return to this point later.

Mode of action Kerr & Htay (1974) soon followed the initial report of control with a likely explanation of the mechanism. Strain 84 was tested against 44 strains of *A. tumefaciens* in tomato seedlings: 34 of them were controlled, one was partly controlled and 9 were unaffected by the control agent. The same strains were then tested against strain 84 on agar: all 34 strains controlled in seedlings were inhibited by a diffusate of strain 84 in agar, the one partly controlled strain was partly inhibited and 8 of the 9 non-controlled strains were unaffected by strain 84 on agar. Inhibition was shown to be due to a bacteriocin termed *agrocin 84* produced by the control agent. The one exceptional strain that behaved differently in the plant test and on agar was shown to produce a bacteriocin of its own that inhibited the control agent.

Bacteriocins are a special type of antibiotic that affect only organisms closely related to the producer organism. The best known examples are the *colicins* produced by *Escherichia coli*; these are proteins with several diverse effects on other strains of *E. coli*. Agrocin 84 is entirely different from any previously described bacteriocin because it is a nucleotide derivative of about 1000 molecular weight, as shown below, and it inhibits DNA synthesis, possibly in a similar way to that of adenosine and its derivatives.

The impressive correlation obtained by Kerr & Htay and mentioned above leaves little doubt that agrocin 84 plays a central role in the control process, but some important points still need to be resolved. For example strain 84 is effective only if its cells outnumber those of the pathogen at wound sites, and this suggests that competition for infection sites plays some part in the control process. Also, other bacteriocin-producing strains of *A. radiobacter* have been isolated and shown to be inhibitory in culture and yet none has proved as successful as strain 84 for disease control. The reasons for these anomalies are largely unknown; it is hoped that further study will lead to the discovery of other effective strains – perhaps for cases in which control by strain 84 has broken down.

Breakdown of control? Fortuitously, the Ti plasmid codes for both tumour production and sensitivity to the bacteriocin, so strains of *A. tumefaciens* that spontaneously become resistant to the bacteriocin are also usually avirulent. This is not invariably true, however, and there are an increasing number of reports from laboratory work of virulent but bacteriocin-resistant mutants. Sule & Kado (1980), for example, detected plasmid mutations that conferred resistance, and the likely reason for resistance in this case was failure of the mutants to take up the bacteriocin. It would be premature to call this a breakdown of control because there is no evidence on whether or not the strains were still subject to control by strain 84 on plants.

A potentially more serious situation arose in Greece in experimental work on control of crown gall in almond seedlings. The expected degree of control was not found and the reason was that the pathogen itself had developed the ability to produce agrocin 84 (a bacteriocin-producer is resistant to its own bacteriocin). In other words, genetic recombination had occurred between the pathogen and the control agent. This is possible because strain 84 contains two plasmids, one of

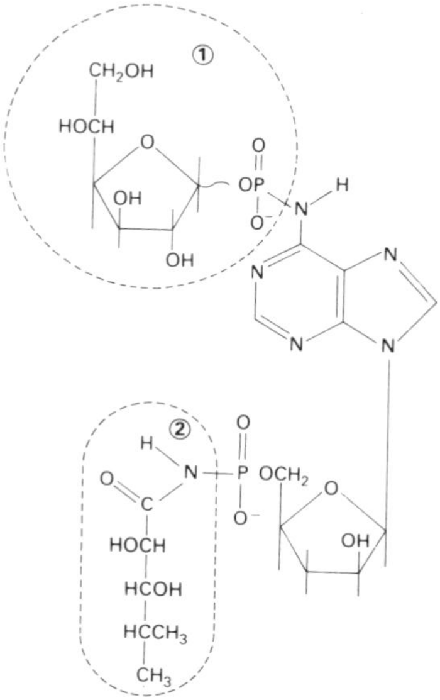

Agrocin 84

Substituent **1** determines activity as a bacteriocin (without it the molecule is a simple antibiotic)

Substituent **2** confers antibiotic activity

which (the larger) codes for agrocin production and is non-conjugative, while the second one is conjugative and, like all conjugative plasmids, can 'mobilize' a non-conjugative one. Thus in appropriate conditions strain 84 can transfer both of its plasmids to the pathogen which is then no longer subject to control. A solution to this problem is being sought by selecting mutants of strain 84 with a defective conjugation or mobilization system. It remains to be seen whether this will be a permanent solution or whether we are witnessing the first steps on a molecular biological treadmill.

Cross-protection against viruses

Cross-protection was first discovered by McKinney in 1929. If a plant is *systemically* (i.e. generally) infected by one strain of a virus it will not develop symptoms when inoculated with a second (challenge) strain. By definition this is a *reciprocal* phenomenon, i.e. it can operate in either direction, and it has proved useful in typing viruses because it is shown only by closely related ones. It differs in this respect from *acquired systemic resistance*, which is a one-way protection shown by unrelated viruses or other pathogens. The idea that it could be used as a control measure, by purposefully inoculating 'mild' or attenuated strains of viruses into plants, developed in the 1950s. Since then it has been used commercially in at least two cases: for control of *tobacco mosaic virus* (TMV) in tomato crops and of the *tristeza* virus of citrus crops in South America.

TMV is one of the most important pathogens of glasshouse-grown tomatoes, causing yield losses of 10 to 15% or more if present early in the growing season but after the first truss of fruit has formed. It causes mosaic of the leaves, stunting of the shoots, and reductions in the number, size and quality of fruit. The traditional approach to control has been by sanitation to remove all cources of inoculum; more recently, multiple resistance genes have been incorporated into some commercial tomato cultivars. These measures are generally effective, but the virus is transmitted so easily from sources of inoculum on clothing, tools and workers' hands that a crop is always at risk.

In 1964 British growers began a deliberate and remarkable policy of infecting plants with fully virulent strains of TMV early in the growing season, because economic losses are much lower if the plants are infected before the first truss of fruit has formed. Widespread use of a microbial control measure came only later, in 1972, when Rast in the Netherlands produced an almost symptomless mutant of the virus by nitrous acid treatment of a virulent strain. The mutant was termed *MII-16* and was subsequently used by up to a quarter of growers in different parts of the world. The mutant itself causes approximately 5% yield loss but this is difficult to assess in practice because plants that are unprotected and do not contain resistance genes almost inevitably become infected by virulent strains at some point in their growth. Cross-protection therefore often gives economic returns, especially if the virulent strains are likely to infect early in the growing season, as shown in Table 5. The practical application of the method is very simple. The mild strain is propagated in tobacco plants which are then ground to obtain the sap. The virus is purified and concentrated by centrifugation; it is then diluted in water, carburundum is added as an abrasive, and the mixture is sprayed on to trays of seedlings. The virus infects through minor wounds caused by the pressure of the spray and, if necessary, by lightly brushing a piece of cardboard across the top of the seedlings. Up to 30 000 seedlings can be treated in one hour, at negligible cost, and no further treatment is required.

Cross-protection of citrus plants against the tristeza virus was begun experimentally in Brazil in 1961, using naturally occurring mild strains obtained from healthy plants in otherwise severely damaged plantations (Costa & Muller, 1980). The first commercial trials were begun in 1968 when some trees artificially infected by mild strains were released to a few growers. The method showed such promise that demand grew rapidly and the growers themselves began to propagate material from the trees originally supplied. This is possible because the virus is systemic and fruit trees are normally propagated by grafting selected buds or scions on to rootstocks. By 1980, 8 million cross-protected trees of Peru sweet orange were being grown commercially. The method has proved effective despite the naturally high level of challenge inoculum (wild virus strains) in the area and the similarly high incidence of aphid vectors of the virus. It is less effective if buds or scions containing the virulent strains are grafted on to cross-protected plants (for experimental purposes – this would not be done in practice), presumably because much higher levels of challenge inoculum are present in the buds and scions than would normally be transmitted by vectors.

The potential use of cross-protection is being investigated in several other cases. One interesting example concerns the *apple mosaic virus* in New Zealand, which can be particularly damaging to susceptible apple cultivars like 'Jonathan'. The replacement of infected trees by healthy ones is expensive, so initial experiments were aimed at revitalizing severely diseased trees by grafting on to them scions containing mild virus strains (Chamberlain *et al.*, 1964). This worked well and looked promising as a control measure, until it was recognized that the virus is seldom transmitted by aphid vectors as had been thought. Rather, the very high incidence of the disease in New Zealand seems to have been due to the use of

Table 5 Yield of tomatoes from plants inoculated with attenuated or wild strains of tobacco mosaic virus and from uninoculated controls grown in simulated commercial conditions in Britain (from Channon *et al.* (1978) by kind permission of the authors and the publisher)

	Fruit yield (kg/plant) and commercial value (£/plant)					
	1973		1974		1975	
	Yield	Value	Yield	Value	Yield	Value
Control (uninoculated)*	7.28	2.01	7.24	2.16	6.86	2.80
Attenuated mutant (MII–16)	6.88	1.92	7.92	2.41	7.26	2.90
Virulent wild type	6.28	1.71	6.95	2.05	–	–

* *The controls developed natural infection at about 14 wk (1973), 7 wk (1974) and 8 wk (1975). In the latter two years use of the avirulent mutant gave an economic return.*

virus-infected material initially for propagation. This means that the problem can be solved in the long term merely by replanting with uninfected material, and this is more attractive economically than cross-protection because the mild strains do cause some yield reduction.

Mechanism of cross-protection The biochemical or molecular basis of cross-protection is poorly understood. The protecting virus must be present in the tissues for the phenomenon to occur because protection is lost if plants are cured of infection by heat therapy. In some cases the challenge inoculum is known to replicate to a limited degree in the host cells whereas in other cases its replication is blocked at an early stage. This might possibly be explained in terms of the *replicase* system. Viruses like TMV have a single-stranded RNA genome which is thought to act as a messenger RNA in the early stages of infection. It acts as a template for the production of a replicase sub-unit (a polypeptide) which then combines with host proteins to form a functional replicase. This attaches to the viral RNA and replicates it as part of the production of new virus particles. Replicases are known to have some specificity so it is possible that the replicase of a mild strain can attach to the RNA of a newly introduced virulent strain (because of a similar binding site) but be unable to replicate it, thereby blocking further development.

Application of specific control agents: general comments

The examples discussed so far differ in detail but have several features in common which largely explain their success. First, the control agent is introduced into a *sterile and selective environment* in which nutrients are readily available to it (in host cells in the case of a virus and in fresh wound tissue in the other cases). Second, each control agent is *ecologically related* to the pathogen against which it is used – it occupies the same site and uses the same nutrient sources. Third, the control agent is applied to a *small and often specific region* of the plant where it is needed, so the treatment is not wasteful. Fourth, the control agent does not need the pathogen to be present in order to maintain itself. Fifth, each example concerns a *high value crop*, a fact that justifies the expenditure of time and money in protecting individual plants.

The disadvantages of these control agents are similar to those mentioned for insect pathogens. In particular, the production, storage and application of living organisms present more difficulties than in the case of chemical control agents, and the timing of application is often critical. Most important, however, is the fact that the control methods considered so far are largely preventative rather than curative – the need for them must be predicted before damage occurs.

It is generally accepted that the addition of large amounts of inoculum of a control agent to soil or even to the leaves of plants is impracticable. Apart from the cost, this approach would not introduce the control agent into the specific site where it is needed (and certainly not at a high enough inoculum level). A further problem is that any organism introduced in this way would need to compete with the resident micro-organisms which inevitably are highly adapted to the prevailing conditions. A more realistic approach is to alter the environment in some way to encourage the activities of selected resident microbes or introduced ones; also, if possible, the existing microbial community should be simplified so that competition is less intense. These are among the approaches discussed in the next chapter. There are, however, further prospects for wound inoculation (including stump protection) and also perhaps for seed inoculation, as discussed below.

Prospects for wound inoculation

At least three examples of wound inoculation are currently showing promise at the experimental stage (Corke, 1978).

1 Apple canker is caused by the fungus *Nectria galligena*, which infects from spores splash-dispersed on to leaf scars formed during leaf fall or on to pruning wounds. Fungicidal sprays are effective in protecting leaf scars but not pruning wounds, so attempts have been made to protect the latter with antagonistic micro-organisms. One of the most effective is the fungus *Trichothecium roseum*, and yet this does not affect *Nectria* in laboratory culture. The reason for its effectiveness is that it promotes secondary colonization of wounds by *Trichoderma viride*, a highly antagonistic fungus that probably brings about control. More work is needed on the ecology of pruning wounds before this approach can be used successfully and predictably, but the example illustrates that even a fungus with no direct effect on a pathogen can sometimes be useful because of its secondary, indirect effects.

2 *Silver leaf* disease of plum and other fruit trees is caused by the fungus *Stereum purpureum*. This enters through wounds, colonizes the woody tissues and produces toxins which are transported to the leaves where they cause the epidermis to separate from the underlying tissues, giving the typical silver leaf symptoms. It has been known since the 1920s that saprophytic fungi can colonize wounds and exclude *Stereum*, but attempts to use these fungi as control agents have been made only recently. *Trichoderma viride* is currently being tested on a pilot scale and it has just received clearance from the registering authorities in Britain for this purpose. The fungus is considered further in chapter 7.

3 The fungus *Eutypa armeniaceae* causes *dieback* and *gummosis* of apricot trees when it gains entry into the plant's vascular system through pruning wounds. Experimental work in Australia has shown that applying fungicides like *benomyl* to trees 5 weeks before pruning (for control of other diseases) significantly *increases* infection by *Eutypa*. The reason is that plant surfaces normally support a population of saprophytic and weakly parasitic micro-organisms (chapter 6) which compete with *Eutypa* for pruning wounds, and these natural control agents are suppressed or eliminated by the fungicide. *Fusarium lateritium* was identified as one of the most important antagonists in this case. Interestingly, *Eutypa* itself is susceptible to benomyl, and pruning wounds can be protected by applying this fungicide; but only short-term protection is achieved in this way because the fungicide levels soon decline. *F. lateritium* is about 10 times less susceptible to the fungicide than is *Eutypa*, so an integrated control programme is being considered in which a mixture of benomyl and *Fusarium* spores is used; the fungicide gives short-term protection until the *Fusarium* can become established and give complete, long-term protection.

An interesting by-product of the research into these examples of microbial control has been the development of modified pruning shears designed to dispense automatically a quantity of fungicide or suspension of a control agent as the cut is being made. Thus, even if the microbial control agents in these cases never reach the stage of commercial exploitation, still the work will have been worthwhile.

Prospects for seed inoculation

The techniques of seed inoculation have been developed over many years and are now used routinely to establish *Rhizobium* on legumes and to establish 'bacterial fertilizers' (mixtures of *Azotobacter chroococcum* and *Bacillus megaterium*) on various crops in the Soviet Union. The same techniques have been used for control of crown gall, so we can ask if there are further prospects for the use of seed inoculation in disease control.

The main types of disease that can be controlled by seed inoculation are the seedling diseases caused by species of *Pythium*, *Fusarium*, *Rhizoctonia* and *Phytophthora* etc. These fungi attack a wide range of plants in the seedling stage but generally are unable to overcome the resistance of mature plants. They are stimulated to grow by the substantial amounts of nutrients released from germinating seeds and from the growing root tips, and only short-term control is needed because within 1 or 2 weeks from germination the plant has developed its normal resistance mechanisms. Numerous laboratory and glasshouse studies have shown that seed inoculation is effective against such diseases, the inoculants being either specially chosen antagonistic micro-organisms (e.g. *Trichoderma*, *Bacillus* spp.) or merely fast-growing competing organisms that can 'mop up' the available nutrients as they are released from the plants. However, fungicides can be equally effective because they are needed for only a short time, and they have the advantage (apart from ease of application, storage etc.) that they can control some pathogens present inside the seed coat whereas microbial control agents are effective only against exogenous pathogens. Many types of seed are now routinely protected by broad-spectrum fungicides like the organic mercurials and thiram, with the result that microbial control is unnecessary.

A more pressing need is for long-term control of pathogens that progressively colonize the root system – for example the take-all fungus (see below, pp. 64–7). Fungicides are generally ineffective in these cases, for several reasons: Those applied to seeds have only a limited sphere of influence, too much material is needed for soil drenches, and no fungicide is yet commercially available that can be applied to leaves and will travel down into the roots (the *systemic* ones, like benomyl, move predominantly in the xylem and thus *upwards* from the points of application). Unfortunately, it is difficult to get a seed-applied micro-organism thoroughly to colonize the root system, so this major potential avenue of microbial control has been little exploited. Some relevant points in this respect are outlined below, from a review by Bowen (1979).

1 Saprophytic microbes in the root zone live on nutrients that leak out of the root cells or become available during cell death. The rate of nutrient supply limits the size of the microbial community at any point on the root, because the population cannot increase beyond the stage at which the maintenance energy requirement of the existing cells equals the rate of nutrient supply. It follows that micro-organisms that become established further down a root before the inoculated one can reach this point effectively exclude it by prior use of nutrients.

2 The speed of colonization of roots by an inoculated organism depends on its growth rate and rate of migration. The generation times on roots are often rather long. For example, estimates of 5.2 h for *Pseudomonas* spp. and 39 h for *Bacillus* spp. – both cases much longer than in aerated liquid culture in the laboratory.

3 The migration rates of bacteria on roots are also usually slow. For example *Pseudomonas fluorescens* was found to spread 2.5 cm along sterile roots in 2 days in very moist soil (less than 0.2 bar suction) but hardly at all in soil drier than this. In these respects, a soil of 0.2 bar suction will yield water under a suction of almost exactly 0.2 atmospheres, and most plants can grow down to at least 1 to 5 bar suction. The fungi are much less affected by soil moisture than are the bacteria, and most will grow well down to at least 20 bar suction. However, unlike the bacteria the fungi spread on roots mainly by hyphal growth which is relatively slow (a few millimetres or so every 24 h) and in good conditions the roots of several plants grow faster than this (e.g. 2–7 cm/24 h for wheat roots, though this is exceptionally fast).

The net result of these and other factors is that roots are often colonized afresh just behind their tips from sources of inoculum in the soil, rather than by microbes previously established on the older root regions. Micro-organisms applied to seeds are thus unlikely to colonize the root system extensively unless they can use sources of nutrients unavailable to the normal natural colonizers.

Summary

Addition of specific microbial control agents has proved successful in several cases. The fungus *Peniophora gigantea* is applied to freshly cut pine stumps to prevent them from being colonized by the pathogen *Heterobasidion annosum*. *Peniophora* excludes the pathogen by hyphal interference, a form of contact inhibition. Strain 84 of the non-pathogenic bacterium *Agrobacterium radiobacter* is applied to seeds or roots of many woody plants to prevent infection by the crown gall bacterium, *Agrobacterium tumefaciens*. This control measure operates through production of a bacteriocin by strain 84. Infection of plants by virulent strains of viruses can be prevented by prior inoculation with mild strains. This phenomenon of cross-protection has been used to control tobacco mosaic virus on tomato plants and the tristeza virus on citrus trees in South America. In all these cases the control agent is applied to a localized part of the plant at low cost and the plant provides a selective environment for its growth. The prospects for seed inoculation seem less good in general because of the difficulties of colonization of a root system from an inoculum applied to seeds.

References

Bowen, G. D. (1979). Integrated and experimental approaches to the study of growth of organisms around roots. In *Soil-Borne Plant Pathogens* (Ed. B. Schippers and W. Gams). Academic Press, London.

Chamberlain, E. E., Atkinson, J. D. and Hunter, J. A. (1964). Cross-protection between strains of apple mosaic virus. *New Zealand Journal of Agricultural Research* 7: 480–90.

Channon, A. G., Cheffins, N. J., Hitchon, G. M. and Barker, J. (1978). The effect of inoculation with an attenuated mutant strain of tobacco mosaic virus on the growth and yield of early glasshouse tomato crops. *Annals of Applied Biology* 88: 121–9.

Corke, A. T. K. (1978). Microbial antagonisms affecting tree diseases. *Annals of Applied Biology* 89: 89–93.

Costa, A. S. and Muller, G. W. (1980). Tristeza control by cross protection: a U.S.–Brazil cooperative success. *Plant Disease* 64: 538–41.

Hardy, K. (1981). Bacterial Plasmids. *Aspects of Microbiology* 4 (Ed. J. A. Cole and C. J. Knowles). Nelson, Walton-on-Thames.

Ikediugwu, F. E. O. (1976). The interface in hyphal interference by *Peniophora gigantea* against *Heterobasidion annosum*. *Transactions of the British Mycological Society* 66: 291–6.

Ikediugwu, F. E. O. and Webster, J. (1970). Antagonism between *Coprinus heptemerus* and other coprophilous fungi. *Transactions of the British Mycological Society* 54: 181–204.

Kerr, A. and Htay, K. (1974). Biological control of crown gall through bacteriocin production. *Physiological Plant Pathology* 4: 37–44.

Moore, L. W. (1979). Practical use and success of *Agrobacterium radiobacter* strain 84 for crown gall control. In *Soil-Borne Plant Pathogens* (Ed. B. Schippers and W. Gams). Academic Press, London.

Rishbeth, J. (1963). Stump protection against *Fomes annosus*. III. Inoculation with *Peniophora gigantea*. *Annals of Applied Biology* 52: 63–77.

Süle, S. and Kado, C. I. (1980). Agrocin resistance in virulent derivatives of *Agrobacterium tumefaciens* harboring the pTi plasmid. *Physiological Plant Pathology* 17: 347–56.

6 Disease control: manipulation of the microbial balance

This chapter and the next are concerned with naturally occurring microbial control agents. In some cases these operate effectively without man's intervention – the subject of the next chapter; in other cases which are considered here their numbers or activities must be increased by appropriate manipulation of the environment. Like the pathogens themselves, these control agents form part of the complex microbial balance in natural situations, so our ability to manipulate them depends on our understanding of the factors that affect this balance.

Four types of manipulation are considered here as representative of the many possible approaches: soil sterilization, use of fungicides, use of soil supplements like crop residues, and crop rotation. In the future it may be possible to breed plants specifically to favour microbial control agents, so this also is briefly discussed. All these procedures usually have direct effects on pathogens as well as indirect effects operating through the activities of microbial control agents. So they are examples of integrated control rather than microbial control in the strict sense.

Soil sterilization

Soil sterilization is almost universally practised in glasshouses, where it is justified by the high value of the crops and where it is necessary because, in the absence of crop rotations, pathogens often build up to damaging levels. It can be achieved by heating or by use of chemical fumigants and in both cases there is an element of microbial control.

Heat treatment Soil is commonly sterilized by steaming at 100°C for 30 minutes. The treatment is highly effective in killing pathogens, pests and weeds, and crop growth can be stimulated initially because of the mineral nutrients released from dead organisms. But the treatment is expensive, is hazardous to workers and has several further drawbacks. The nitrifying bacteria (*Nitrobacter*, *Nitrosomonas*) usually take longer to become re-established than do the ammonifiers (a wide range of saprophytes that release ammonia from organic nitrogen sources) so in some soils ammonia can accumulate to toxic levels. Drastic heating can release toxic levels of water-soluble manganese from the soil organic matter, as well as several phytotoxic breakdown products of the organic matter itself. Most importantly from our present viewpoint, the treatment is unnecessarily harsh because it destroys soil saprophytes as well as parasites, and some of the saprophytes are microbial control agents. Heated soil is rapidly recolonized by micro-organisms that use the organic nutrients made available during sterilization, and some of these recolonizers are fast-growing pathogenic fungi like *Pythium*, *Fusarium* and *Rhizoctonia* spp. which cause *damping off* diseases of seedlings. The result can be higher levels of some diseases than if the soil were left unsterilized.

A practical solution to this problem was devised by Morris in Britain in 1954 but was first applied in the 1960s in California. Milder heating can be achieved by using steam: air mixtures (*aerated steam*) whose temperatures can be adjusted at will by altering the ratio of steam to air (Baker, 1970). Most pathogens are killed by exposure to steam at 60°C for 30 minutes, and this, with some margin for error, is the treatment now commonly used in practice. It does not eliminate the saprophytic microflora; indeed the spore-forming bacteria like *Bacillus* spp. usually increase in numbers after treatment because of the newly available substrates, the reduced competition and the fact that their spore dormancy is broken by heat. This residual population prevents the soil from being recolonized by any pathogens that persist in local pockets or around the treated area.

The effectiveness of this treatment can be seen in Table 6, from the work of Olsen & Baker (1968). A series of seed trays was filled with natural soil heat-treated to different extents and then a thick crop of pepper seedlings was grown. The pathogen *Rhizoctonia solani* was introduced into one corner of each tray (to simulate a surviving pocket of inoculum) and the spread of disease from this inoculum was assessed. In untreated soil some of the seedlings died in local pockets owing to infection by a *Pythium* sp. naturally present in the soil. All heat treatments killed the *Pythium*, but Table 6 shows that steaming at 100°C left the soil completely unprotected against subsequent invasion by *Rhizoctonia* whereas lower temperatures gave progressively more protection, 60°C being best. Microscopical examination showed that this protection was due to *Bacillus subtilis* which colonized the *Rhizoctonia* hyphae emerging from the inoculum, causing them to lyse.

Table 6 Extent of disease caused by *Rhizoctonia solani* in a crop of pepper seedlings grown in soil treated for 30 minutes at different temperatures. (From Olsen & Baker, 1968)

Temperature (°C)	Disease caused by *R. solani*		Area of disease caused by resident *Pythium* sp.
	Area (cm^2)	Linear spread (cm)	
(None)	–	0.3	102.6
100	253.4	17.8	0
71	64.5	8.8	0
60	3.2	2.0	0

The use of steam is almost entirely restricted to glasshouse conditions, but recent research in Israel and the USA has shown the feasibility of using solar heating to eliminate pathogens in the field. This is done by laying sheets of thin transparent polythene on the soil for 2 to 4 weeks; the temperature is raised by 10°C or more in this way, even down to 30 to 40 cm soil depth. Evidently the intermittent mild heating is enough to kill some plant pathogens; it probably acts by a combination of direct and indirect effects, the latter operating through the activities of microbial control agents. The method shows promise for commercial use, and reviews of experimental work on it can be found in Schippers & Gams (1979).

Soil fumigation Treatment of soil with chemical fumigants is done commercially for high-value 'market garden' crops and in forest nurseries, etc. but is generally too expensive for most field crops. The fumigants commonly used include *methyl bromide*, *chloropicrin*, *Dazomet* (which breaks down to yield ultimately *methyl isothiocyanate*) and *dichloropropane-dichloropropene* mixtures. They have different rates of spread through soil as vapours and this is the main basis on which a choice is made between them; but they also have somewhat different spectra of activity against the main groups of micro-organisms, so in theory they could be chosen for their partly selective effects. A good example of a selective effect is seen in the use of *carbon disulphide* (CS_2) to control *Armillariella mellea* in Californian peach orchards.

Armillariella has a world-wide distribution as a pathogen of trees and is particularly damaging to young trees planted on previously afforested sites. For example it occurs in tea, rubber and cocoa plantations in tropical and sub-tropical countries where the indigenous forest is cleared to plant these crops. It is a problem also in several orchard crops in the USA and it is perhaps the second most important pathogen (after *Heterobasidion annosum*) in British forests, where it kills young conifers planted on previous 'hardwood' sites. The fungus causes butt rot and root disease, but it is basically only a weak parasite that needs a substantial inoculum base from which to grow and infect new plants, and this is provided by the stumps and major roots of previously damaged trees. Control measures are aimed at reducing this inoculum base.

A standard control measure in Californian peach orchards is to remove the stumps of diseased trees and to use CS_2, a weak fumigant, to kill the fungus in the roots left after stump extraction. Initially CS_2 was thought to kill the pathogen outright, but this view was questioned when work in the 1950s showed that fumigation with CS_2 does not kill the fungus in roots held in sterile conditions. Burial in soil is necessary, and it was seen that other fungi, especially *Trichoderma* spp., normally replace the pathogen in roots after fumigation. Yet *Trichoderma* alone seldom replaced the pathogen in roots that had not been fumigated, so it seemed that a combination of fumigant action and *Trichoderma* is necessary to explain the success of CS_2 in practice.

Trichoderma spp. frequently become dominant in fumigated soils and yet they are not particularly resistant to the fumigants (CS_2, methyl bromide, etc.). Their success seems to be due to the facts that (1) they are fast-growing (e.g. about 25 mm/24 h across agar plates at 25°C) so they can recolonize rapidly, (2) they are more resistant to fumigants than are even faster growing fungi, and (3) they recover quickly from exposure to sub-lethal doses of fumigant such as they are likely to encounter at the margins of a treated area. In contrast, *Armillariella* is very sensitive to fumigants and takes several days to recover from exposure to sub-lethal doses. Combining all these points, CS_2 is thought to act as follows. Some of the *Armillariella* is killed outright by CS_2 but some, because it is protected by being inside the root tissues, is exposed to sub-lethal doses and is 'weakened'. *Trichoderma* spp. rapidly recolonize the fumigated soil, invade the roots because *Armillariella* is weakened, and kill the pathogen. The mechanism of the weakening was subsequently explained by Ohr & Munnecke (1974) who showed that *Armillariella* normally produces antibiotics which prevent other fungi from invading the roots, but after exposure to sub-lethal doses of fumigant the production of antibiotics is reduced or stopped for several days. This still leaves unanswered the question of how *Trichoderma* kills the pathogen. In culture *Trichoderma* spp. frequently coil around and penetrate the hyphae of other fungi, indicating *mycoparasitism* (Figure 13), but several other explanations are possible, as discussed below (pp. 73–4).

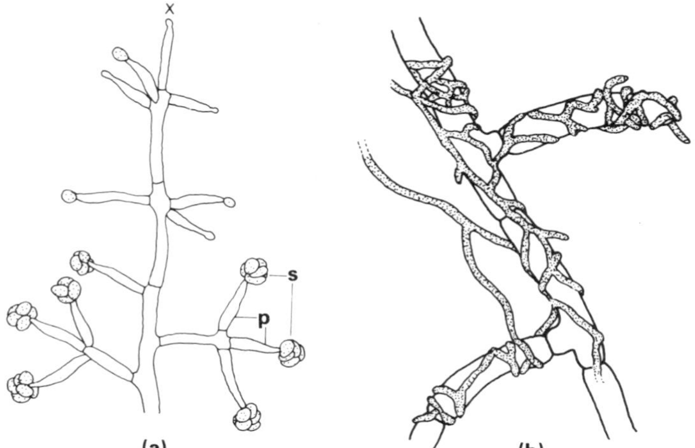

Fig. 13 *Trichoderma* spp. **a** *Trichoderma viride*, showing typical arrangement of spores and phialides (s, p); in many other *Trichoderma* spp. the same arrangement is seen but the phialides are shorter and the spore-bearing hyphae may be extended into sterile appendages at the point marked 'X'. **b** Coiling of hyphae of *Trichoderma* around the wide hyphae of another fungus *Rhizoctonia solani*.

Microbial control by use of fungicides

Fungicides differ from fumigants in several respects. They do not normally volatilize, they are more selective in toxicity, and relatively few are applied to soil except on or around the seeds – instead they are used mainly to control diseases of the above-ground parts of plants. Like fumigants, they are used for their direct effects on 'target' organisms but almost inevitably they also affect 'non-target' species. This can be detrimental or beneficial, depending on circumstances. For example, returning to the *Armillariella*/*Trichoderma* interaction, the use of benomyl actually increases disease in experimental conditions, because *Armillariella* is resistant to this fungicide (like most of the Basidiomycotina) whereas *Trichoderma* is susceptible to it (like many of the Deuteromycotina). A similar detrimental effect of this fungicide was seen in the case of *Eutypa* on apricots in chapter 5 (p. 55). In other cases fungicides may help to promote microbial control, and one such possible case is considered in detail below to illustrate how the effect might operate.

In New Zealand and parts of Australia sheep sometimes develop *facial eczema* disease when they ingest spores of the fungus *Pithomyces chartarum* whilst grazing on infested pasture. The spores contain a hepatotoxin termed *sporidesmin*, which is absorbed in the gut and causes extensive necrosis of the liver and bile duct, the latter becoming blocked by scar tissue. Normally the chlorophyll absorbed by the gut is converted to phylloerythrin in the liver and this re-enters the gut via the bile duct. In toxin-affected sheep, however, the blockage of the bile duct causes phylloerythrin to accumulate in the blood and, being photoactive, it causes blisters and sores to develop on parts of the skin exposed to sunlight – hence the name facial eczema.

Pithomyces is a saprophyte that grows on the dead leaves of grasses and other pasture plants. It is particularly common in late-summer following drought or serious infection by leaf pathogens, i.e. anything that contributes to the amount of dead herbage. In suitably humid conditions the fungus then sporulates and the sheep are thus exposed to the toxin. One recommended control measure has been to spray pastures at intervals during the summer with the fungicide *benomyl*. As shown in Figure 14, this causes a marked reduction in the number of spores, and it is a direct antifungal effect because the growth of *Pithomyces* in laboratory culture is prevented by only 0.5 ppm (μg/ml) benomyl. But benomyl also affects the non-target species in New Zealand pastures: sporulation by *Cladosporium* spp. is markedly suppressed, whereas that by *Alternaria* and *Stemphylium* spp., for example, is equally markedly increased after fungicide treatment (Figure 14). The latter two fungi are resistant to benomyl in culture, their growth being unaffected by even 100 ppm. The significance of this for the control process is seen by considering the ecologies of all these fungi, as outlined below.

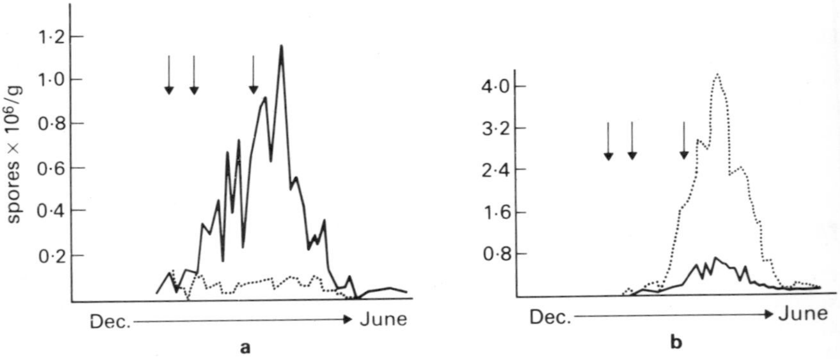

Fig. 14 Numbers of spores of **a** *Pithomyces chartarum* and **b** *Alternaria tenuis* in New Zealand pasture in the course of a season. Solid line, untreated pasture; broken line, pasture sprayed with the fungicide benomyl at three times marked by arrows. (Taken from McKenzie (1971) by kind permission of the author and the publisher.)

Most plants support a characteristic *leaf surface microflora* of saprophytes and weak parasites which grow on the small amounts of nutrients that leak out of the leaves or are present in aphid honeydew or even in material like pollen grains deposited on the plant surface. In temperate climates this microflora includes *Cladosporium* spp., *Alternaria* spp., 'pink yeasts' (e.g. *Sporobolomyces*), 'white yeasts' (e.g. *Cryptococcus*) and chromogenic bacteria. The activities of all these organisms increase as the leaf senesces, but they are replaced by other species (*Helminthosporium*, *Stemphylium*, and *Pithomyces* in appropriate climatic conditions) as the leaf dies and becomes part of the plant litter. Now, it is reasonable to assume that each species has particular attributes that enable it to live as part of this microbial community; it is equally reasonable to assume that their substrate requirements overlap to some degree, because all can utilize the sugars and simpler plant cell wall polymers as energy sources. So, if some species are prevented from growing by benomyl treatment, others are likely to encroach on their 'territory' or delay their re-establishment when the fungicide level declines. In other words, the direct effect of benomyl on *Pithomyces* is probably reinforced by a subsequent microbial control. In support of this suggestion, Figure 14 shows that *Alternaria* clearly benefits from the suppression of *Pithomyces* and other fungicide-susceptible species, indicating that it can use the nutrients previously used by them. This is not an isolated example; in fact there is a large body of information on the interactions of leaf surface saprophytes with plant pathogens, summarized in several of the papers in Dickinson & Preece (1976). The problem with such interactions on the surface of living plant parts (unlike the situation in leaf litter mentioned above) is that the amounts of available nutrients are low and thus the scope for significant manipulation of the microflora is strictly limited.

Use of soil supplements

Organic supplements include materials like crop residues, farmyard manure and specially grown *cover crops* – particularly legumes – which are ploughed into the soil. Traditionally they have been used to maintain or improve soil structure and fertility, but they have long been known to have an additional effect in reducing the inoculum of soil-borne plant pathogens during their resting or saprophytic stages. Here we consider them mainly because they involve different mechanisms of microbial control from those that we have seen so far.

In most soils for most of the time there is a chronic energy shortage, all other factors being suitable for microbial growth. Thus the most obvious response to the addition of readily decomposable organic matter to soil is an increase in the activity of micro-organisms, and this is usually a non-specific response involving all components of the microbial community. It results in an increase in the numbers of competitors and antagonists of pathogens, but the mechanism of disease control goes deeper than this. The spores of most fungi – whether saprophytes or parasites – do not germinate in soil even though in many cases they would germinate in distilled water. They are said to be subject to *fungistasis* (or *bacteriostasis* in the case of bacterial cells) and this is an exogenously imposed dormancy quite separate from the endogenous or constitutive dormancy shown by some fungal and bacterial spores. It is caused either by germination inhibitors in the soil or by the continuous withdrawal of nutrients from the spore and its immediate surroundings, but in either case it can be overcome by adding nutrients. Some organic supplements are thought to act by stimulating spore germination by pathogens, and the emerging hyphae must then compete for nutrients with an enhanced population of saprophytes. As most plant pathogens are very poor competitive saprophytes their hyphae are subjected to nutrient-stress and they lyse. The sequence is termed *germination-lysis* and it is one of the main ways in which organic supplements bring about microbial control.

Lysis is caused by enzymes, and a central question is whether these are produced by the pathogen itself or by soil saprophytes (autolysis or heterolysis as explained on p. 43). The problem is far from completely resolved but Ko & Lockwood (1970) in a classic paper showed that germlings added to soil lysed more quickly if they were initially alive than if they had been heat-killed. Taking such evidence at face value, it seems that autolysis is of major significance in soil. This leads to the distinction between competition and antagonism, as mentioned on p. 43. In the laboratory autolysis can be shown to result from nutrient-stress and therefore could be induced by competition, whereas heterolysis can be considered a form of antagonism.

Unfortunately, the use of organic supplements is a double-edged weapon. Some of the least specialized pathogens, like *Rhizoctonia solani* and *Corticium rolfsii* which cause seedling diseases, can compete effectively for organic materials in soil and thus increase their inoculum levels. This is especially true if the organic materials are still living when added to soil because their residual host resistance favours parasites over saprophytes. In other cases parasitic fungi like the *Fusarium* spp. can respond to nutrient-stress by rapidly producing thick-walled resting spores (*chlamydospores*) from some of their hyphal compartments. This process involves autolysis and the remobilization of nutrients from the hyphae into the spores, which are directly equivalent to bacterial endospores. Consequently, these fungi can form one or more resting spores after germination of the original spore (depending on how much growth has been made in the intervening period) and the new spores can have much greater survival value than did the original spores. We see, then, that the effects of adding organic supplements to soil cannot always be predicted, and the method can seldom be advocated without detailed study of individual circumstances.

Crop rotations

The traditional use of crop rotations rests on the fact that most pathogens are host-specific and in the case of those that persist in soil their inoculum levels progressively diminish in the absence of suitable hosts. However, economic pressures have forced increasing specialization on farmers, so that some crops are now grown continuously on the same land (chapter 7) and many are grown *intensively* (in most years), *break crops* being used when pest, disease or weed problems cannot be solved satisfactorily in other ways. For example, a 1 or 2 year break from cereals is needed to reduce the inoculum level of the take-all fungus once this has built up on cereals over a number of seasons; this is the time taken by the fungus to die out from the crop residues that it colonized in its parasitic phase.

In all cases the soil microflora undoubtedly contributes to the gradual decline of inoculum by 'starving out' the pathogen, and in some cases the non-host crops in a rotation play a complementary role. For example they may cause germination-lysis of the spores or other resting structures, because nutrients released from plant roots stimulate the spores to germinate. There is a surprising lack of specificity in this respect; spores usually germinate in response to host or non-host roots, with the result that they often cannot invade the plant and the young hyphae lyse. Plants grown deliberately for this effect are termed 'decoy crops' but it is probably a general phenomenon. In a few cases the break crops in a rotation are known to have an additional and more interesting role because they support populations of specific microbial control agents. For an example of this we return to the take-all fungus, *Gaeumannomyces graminis*, mentioned earlier.

G. graminis is an important pathogen of intensively grown cereals in many parts of the world. It grows along the root surface as darkly pigmented 'runner hyphae'

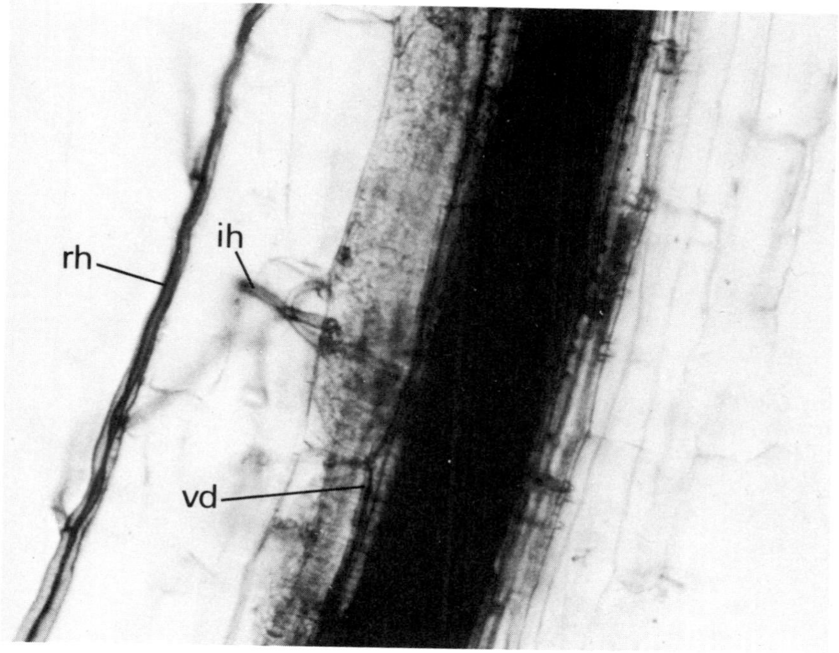

Fig. 15 Wheat root infected by the take-all fungus, *Gaeumannomyces graminis*
rh, runner hypha on root surface; ih, infection hypha; vd, vascular discoloration (blockage).

and from these it penetrates the root cortex by means of hyphal branches which eventually reach the vascular system (Figure 15). The phloem is then destroyed at the point of invasion and the xylem becomes blocked by brown gum-like deposits, with the result that the root ceases to function below that point. If enough roots are infected in the course of a season then the plant dies as a result of water- or nutrient-stress. *G. graminis* can also infect a wide range of grasses in experimental conditions, but at least in Britain it is seldom found on grasses in the field. Instead a similar but non-pathogenic fungus, *Phialophora graminicola*, is found on grass roots. It too grows as dark runner hyphae (Figure 16) and attempts to invade the root cortex by hyphal branches, but its penetration of the cortex is halted by host resistance mechanisms with the result that it never reaches the vascular system and never causes disease.

Both *P. graminicola* and *G. graminis* grow well on cereal or grass roots in glasshouse conditions and they can be shown to interfere with each others' establishment. If *Phialophora* is established first it can help to control take-all, and on this basis it is thought to be responsible for a natural widespread control of take-all in British grasslands (Deacon, 1981). The problem is to transpose this degree of control to cereal crops which, for reasons not fully known, do not naturally support large populations of *Phialophora*. This seems to be achieved if cereals are grown in sequence after grass. The cereals then become infected by *Phialophora* persisting in the soil from the previous grass crop, and even the second and third cereal crops in a sequence can bear significantly more *Phialophora* than if they were grown after other crops. This is not a control measure that originated in the laboratory; on the contrary it has been practised for years in the use of grass in rotations but the central role of *Phialophora* has only recently come to light.

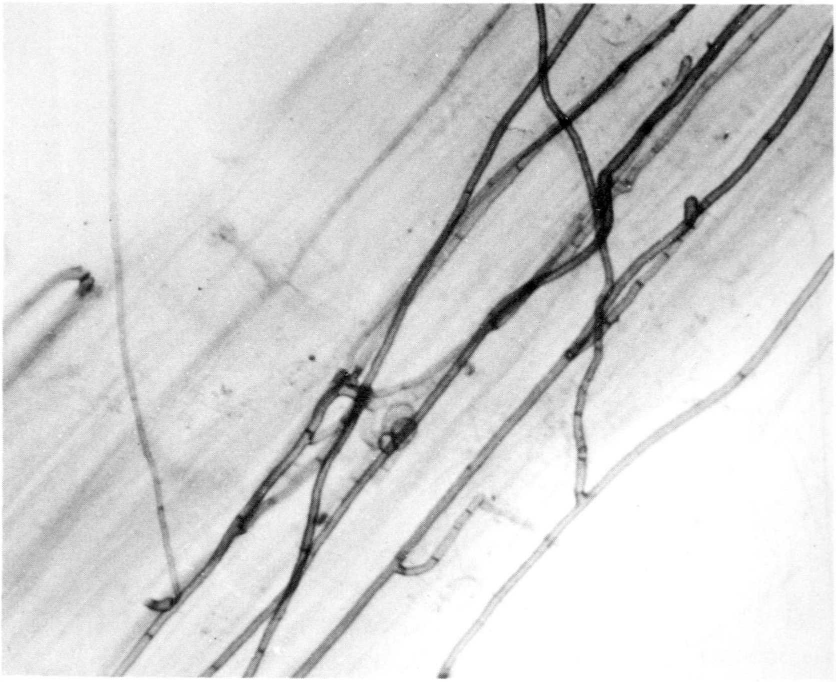

Fig. 16 Wheat root infected by *Phialophora graminicola*, a non-pathogenic parasite. Note the runner hyphae but absence of vascular discoloration.

The mechanism of control in this case is still in doubt. An attractive hypothesis is that *Phialophora* induces a host-resistance mechanism that acts against the take-all fungus. However, no such mechanism has been described – and certainly not one that could explain how the take-all fungus interferes with the establishment of *Phialophora* (the phenomenon is reciprocal). At present competition seems to be the most likely explanation and it is supported by recent evidence on the ageing process in cereal and grass roots. We know that the root cortex of these plants has a limited life and that it dies progressively, starting at the epidermis, with increasing distance behind the root tip. *Phialophora* seems to be a weak parasite, specifically adapted to colonize the cortical cells as they senesce in the normal course of root ageing and before they become colonized by soil saprophytes. In other words, being a parasite it gains an advantage over saprophytes because it can colonize cells with some residual host resistance and in this respect it is similar to other non-pathogenic parasites like *Peniophora gigantea* (see above, p. 46). *G. graminis* is an aggressive pathogen, able to overcome the normal resistance of healthy cells of its hosts, and if it is present at a high enough inoculum level it causes disease irrespective of whether *Phialophora* is present. But from a very low inoculum level, such as is likely to be present in soil after a 1 or 2 year break from cereals, *G. graminis* also probably needs to grow in senescing cells to increase its food base for invasion of the living cells. This would explain why prior colonization of roots by either of these fungi can reduce the establishment of the other – they compete for the same nutrient sources while these are still unavailable to the majority of soil micro-organisms. In many respects these comments about food bases for infection are similar to those made earlier in the case of *Heterobasidion* and *Armillariella* (see above, pp. 45 and 60) and they find parallels in the case of several intestinal pathogens of higher animals.

Manipulation of the microbial balance

General comments Several unifying themes run through the examples in this chapter. (1) The potential for microbial control exists in almost all disease situations; in fact microbial control is probably already operating in many cases but at an ineffective level. Thus it is often unnecessary to add microbial control agents because by suitable manipulation of the environment we can harness the activities of existing ones. (2) The manipulations themselves have much in common. Many of them depend on an initial simplification of the microbial community, followed by the creation of conditions that favour selected members. Both of these ends are achieved by soil heating and by the use of fungicides and fumigants because the substrates previously occupied by the affected species become available for use by the remaining organisms. (3) Because the remaining organisms have been selected from the original community they are likely to be well adapted to the environmental conditions – more so than introduced organisms.

The examples concerning *Armillariella, Pithomyces*, organic supplements and *Phialophora* show some of the complexity of microbial communities and some of the types of interaction that occur between microbes in nature. A glance through the examples also shows that in all cases simple competition could be the main mechanism of microbial control. It is probably the most important of all interactions, because it is the only one that all organisms are likely to be exposed to and that all in turn can exert upon others. It is, however, extremely difficult to demonstrate conclusively.

Future prospects There are some exciting prospects on the horizon for manipulation of the microbial balance, especially in relation to the development of phloem-translocated fungicides and plant cultivars with differential effects on the saprophytic microflora.

Perhaps the most significant advance in fungicide research came in the 1960s when *systemic* fungicides like benomyl were developed. As we have seen, however, these move predominantly upwards with the sap stream so they are largely ineffective against root diseases. Phloem-translocated fungicides are now beginning to be developed; the hope is that they will be applied to leaves and will travel down to the roots and then out to the root surface. If so, they are likely to have special relevance to microbial control because bacteria like *Bacillus* and *Pseudomonas* spp. are probably the most significant antagonists of pathogenic fungi in the root zone, and bacteria, of course, would be least affected by specifically fungicidal agents.

The development of plant cultivars for microbial control is perhaps further off. It first became a reasonable possibility in the 1970s when Canadian workers showed that wheat could be made resistant or susceptible to *common root rot*, a disease caused primarily by the fungus *Cochliobolus sativus*, by disomic chromosome substitution between susceptible and resistant parent cultivars. This in itself is not unusual because the genetics of resistance in plants is already widely exploited. The significance is that the chromosome in question, termed 5B, also determines the number and types of bacteria on the roots. Plants made susceptible to the disease by receiving chromosome 5B from a susceptible parent have twice the bacterial numbers on their roots compared with resistant plants, and 30 to 70 times the numbers of cellulolytic, pectinolytic and amylolytic bacteria. Another chromosome (5D) which is not directly involved in the response to root rot was shown to determine the ability of plants to support growth of free-living nitrogen-fixing bacteria on the roots. None of this is necessarily related to microbial control of disease, but it showed for the first time that the host genotype can be manipulated to favour selected saprophytes in the root zone, and this in turn raises the possibility of selectively encouraging microbial control agents.

Summary

The natural microbial balance can be manipulated in various ways to increase the activities of microbial control agents. Treatment of soil with aerated steam at 60°C controls damping off diseases and also selectively enhances the population of antagonistic *Bacillus* spp. Weak fumigants like carbon disulphide are used against the pathogen *Armillariella mellea* and they operate partly by promoting the activities of antagonistic *Trichoderma* spp. Use of the fungicide benomyl in New Zealand pastures suppresses sporulation of *Pithomyces chartarum*, the cause of facial eczema disease of sheep, whilst stimulating the growth of potentially competing or antagonistic fungi. Soil supplements like crop residues can reduce the inoculum of plant pathogens in soil by stimulating spore germination; the young hyphae are then subjected to intense competition (and antagonism?) from saprophytic micro-organisms and they lyse. Non-host plants in crop rotations can also cause germination-lysis of the spores of pathogens and in some cases the break crops support growth of specific microbial control agents, like *Phialophora graminicola* on grass roots.

Selection of microbial control agents from the existing community has several advantages over the use of introduced organisms. Indeed many treatments that have direct effects against pathogens also selectively encourage microbial control agents and are useful in integrated control programmes.

References

BAKER, K. F. (1970). Selective killing of soil micro-organisms by aerated steam. In *Root Diseases and Soil-Borne Pathogens* (Ed. T. A. Toussoun, R. V. Bega and P. E. Nelson). University of California Press, Berkeley.

DEACON, J. W. (1981) [Take-all] Ecological relationships with other fungi: competitors and hyperparasites. In *Biology and Control of Take-all.* (Ed. M. J. C. Asher and P. J. Shipton). Academic Press, London.

DICKINSON, C. H. and PREECE, T. F. (1976). *Microbiology of Aerial Plant Surfaces.* Academic Press, London.

KO, W. and LOCKWOOD, J. L. (1970). Mechanism of lysis of fungal mycelia in soil. *Phytopathology* 60: 148–54.

MCKENZIE, E. H. C. (1971). Seasonal changes in fungal spore numbers in ryegrass-white clover pasture, and the effects of benomyl on pasture fungi. *New Zealand Journal of Agricultural Research* 14: 379–92.

OHR, H. D. and MUNNECKE, D. E. (1974). Effects of methyl bromide on antibiotic production by *Armillaria mellea. Transactions of the British Mycological Society* 62: 65–72.

OLSEN, C. M. and BAKER, K. F. (1968). Selective heat treatment of soil and its effect on inhibition of *Rhizoctonia solani* by *Bacillus subtilis. Phytopathology* 58: 79–87.

SCHIPPERS, B. and GAMS, W. (1979). *Soil-Borne Plant Pathogens.* Academic Press, London.

7 Disease control: some further mechanisms

The purpose of this chapter is to consider some examples of naturally occurring microbial control, with particular emphasis on mechanisms that have not been dealt with so far. The main subjects for discussion are the spontaneous decline of disease that sometimes occurs during monoculture and the possibly related phenomenon of soils naturally suppressive to pathogens. These are amongst the most valuable examples of microbial control because they operate in field conditions without special manipulations. We shall see the suggested roles of fungal viruses, mycoparasites and antibiotic-producers in these respects and also consider for the first time the diseases caused by plant-parasitic nematodes.

In a different context, we consider the role of mycorrhizal fungi in protecting plants from disease – a role quite separate from their more widely known one in increasing mineral nutrient uptake by plants. Lastly we consider the potential for disease control through induced plant resistance, a topic of much current interest because it relates to the ways in which plants recognize and respond to parasites.

Disease decline and suppression

Figure 17 shows a classic example of *disease decline*. Wheat or barley is often severely infected by the take-all fungus *Gaeumannomyces graminis* (see also pp. 64–7, above) during the early years of monoculture, but after a peak in about the third or fourth year the disease usually declines spontaneously to about half its maximum level. This sequence is accompanied by a trough and then a recovery in crop yields, and cereals can then be grown continuously with acceptable disease levels and reasonable returns. The disease is always present but never in a severe form, and the plants can tolerate this degree of damage provided that they are adequately fertilized. Take-all decline is not an isolated example because similar decline during monoculture has been reported for the cereal cyst nematode *Heterodera avenae*, for potato scab caused by the actinomycete *Streptomyces scabies* and for flax wilt caused by a 'special form' of the fungus *Fusarium oxysporum*.

In contrast to decline, some soils are naturally *suppressive* to particular pathogens, so serious disease never develops during monoculture. For example, some soils in Australian avocado orchards are naturally suppressive to the fungus *Phytophthora cinnamomi*; some soils in Canada, the USA and France are naturally suppressive to forms of *Fusarium oxysporum*, and a soil in Colombia, South America, was recently reported to be naturally suppressive to *Rhizoctonia solani*. Such natural suppressiveness may be widespread but unrecognized simply because pathologists are not asked to investigate healthy crops.

We can consider decline to be *induced suppressiveness* and deal with natural and induced suppressiveness together because they have much in common. (1) They can be demonstrated experimentally by the failure of a soil to support disease when inoculated with a pathogen and sown with a susceptible crop; in this

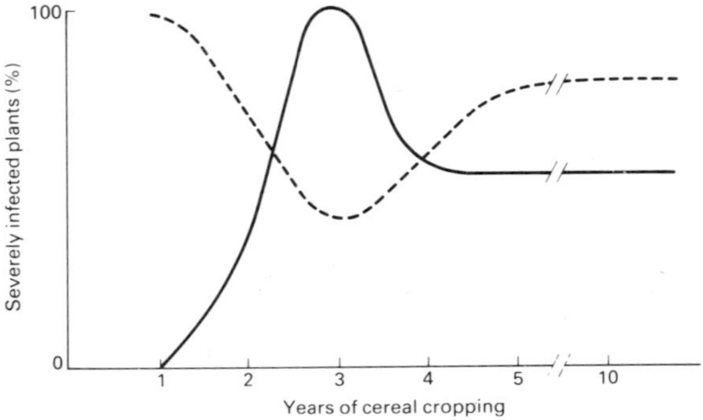

Fig. 17 Take-all decline in continuous cereal crops. Solid line, percentage of plants with severe take-all symptoms; broken line, crop yield.

respect the soils differ from normal *conducive* soils. (2) The suppressiveness is microbial in origin; it is eliminated by biocidal treatments and in appropriate experimental conditions it is transferrable on dilution from a suppressive to a conductive soil. (3) Suppressiveness operates not only against the resident population of a pathogen but also against introduced inoculum, even at quite high levels. (4) In almost all cases the suppressiveness is *specific* to a particular pathogen; a take-all suppressive soil, for example, is not necessarily suppressive to *Rhizoctonia* and vice-versa. In each case, therefore, we are looking for one or more specific micro-organisms that are present only in suppressive soils, or at least are present or active at higher levels in suppressive than in conducive soils.

Take-all decline (TAD) First recorded in Britain in 1933, this has since been reported from Australia, France, Holland, Switzerland, the USA and Yugoslavia. It is the best-known and most widespread example of disease decline and yet still the mechanism is in doubt. TAD takes a few years to develop in the field (Figure 17) but a similar suppressiveness can be induced more quickly in the glasshouse by growing a series of wheat crops for one month at a time and adding inoculum of *G. graminis*. Recently it has been shown that the soil around infected roots is suppressive even in the case of a single one-month old wheat crop grown in partially sterilized soil in a glasshouse. The simplest way of reconciling this with the slow development of suppressiveness in the field is to assume (probably correctly) that suppressiveness does develop quickly around individual roots in the field but takes some time (and several crops) to develop throughout the soil; meanwhile, suppressiveness on a field scale is masked because the pathogen is still spreading from its initial foci of infection within the crop and reducing the yield. In other words, on a field scale we cannot expect to see the gross effects of disease decline until the pathogen has spread through most of the crop. All the evidence suggests that suppressiveness does not develop in healthy crops – the pathogen must be present – but some experimental work shows that it can develop in the presence of the pathogen alone, without the crop.

Most hypotheses about the mechanism of TAD centre around an early report that initially the plants have the same number of infections on their roots in decline and non-decline soils, the difference being that in a decline soil the pathogen does not spread far along the roots from the initial foci of infection. This suggests either that the pathogen itself is deficient or that inhibitory micro-organisms develop on the roots in a decline soil. The first possibility is currently unpopular though it is mentioned later in relation to fungal viruses. The second possibility is favoured, and fluorescent pseudomonads or similar bacteria are suggested to play a major role. Thus, work in the USA and Australia, summarized by Cook & Rovira (1976), has shown that fluorescent pseudomonads are up to 100 to 1000 times more common on take-all diseased than healthy wheat roots, and they were seen to colonize hyphae of *G. graminis* causing these to lyse. They were also up to 100 times more common in suppressive than conducive soil; they could be added to conducive soil which then became suppressive in pot tests, and some of them were shown to antagonize *G. graminis* in culture by producing antibiotics. More recently, the emphasis in take-all suppression has been switched from the root surface to the inoculum units of *G. graminis* in soil (colonized pieces of host debris or artificially inoculated oat grains). Antagonists are thought to operate in these or on their surfaces and to reduce the ability of *G. graminis* to initiate infections of roots. Nevertheless, *Pseudomonas* spp. or similar non-sporing rods are thought to be involved, so this change in emphasis is perhaps not too important. In all cases the pathogen itself clearly promotes the activities of its control agents which then prevent it from initiating infections or spreading from the initial infections.

The problem with this hypothesis is that we cannot explain how *G. graminis* might promote the growth of specific bacteria. Moreover, there is no direct evidence for antibiotic-production by bacteria on roots or in the fungal inoculum in soil. This is not surprising because antibiotics are readily adsorbed on to clay colloids and organic material in soil, and if they are not adsorbed then they are usually broken down by soil micro-organisms. In either case they cannot be detected. Antibiotics may well be produced in significant amounts and may affect *G. graminis* in the immediate vicinity of their production, but the problem is reminiscent of the control of crown gall by *Agrobacterium radiobacter* (see pp. 48–51, above) in which case bacteriocins are implicated but have never been detected in the root zone.

Decline of *Endothia parasitica* (chestnut blight) The fungus *Endothia parasitica* is one of the most damaging plant pathogens; between 1904 and 1950 it virtually eliminated the American chestnut *Castanea dentata* in the USA and thus destroyed one of the most valuable timber crops. It is a wound parasite that causes *cankers* on the twigs, branches and main trunks of chestnut trees and can grow saprophytically (though causing little damage) on a range of other trees like oaks. It kills when the cankers spread around a branch or stem and girdle it.

Endothia has also been a serious problem on sweet chestnut (*Castanea sativa*) in Europe, but in Italy in 1951 it was noticed to be causing less damage in some woodlands than was usual, and by 1978 the disease in Italy had declined naturally to a tolerable level. The explanation was provided by work in France which showed that isolates from trees with sub-lethal damage were markedly less virulent than isolates from girdling cankers. This *hypovirulence* was transferable to fully virulent isolates in paired cultures on agar plates or when mixtures of isolates were inoculated into trees. It was suggested, therefore, that hypovirulence had progressively spread through the population of *Endothia* in Italy, bringing about a natural microbial control. The phenomenon is now being manipulated artificially in France and the USA to cure trees already infected by virulent strains, by inoculating hypovirulent strains at the margins of spreading cankers.

There is overwhelming circumstantial evidence that the *transmissible hypovirulence* is caused by a fungal virus (i.e. a *mycovirus*) or at least by double-stranded RNA of viral origin. Thus, ds-RNA is found in all strains that exhibit transmissible hypovirulence but not in fully virulent strains, and it is always transmitted along with hypovirulence in crosses between strains. Viruses are implicated because the great majority of mycoviruses studied to date contain ds-RNA but, so far, virus-like particles have been seen in only one of many strains of *Endothia* examined. In this case a quite different preparation method was used than is normal, and the virus particles were themselves unusual because they were club-shaped and pleiomorphic in contrast to the normal icosahedral shape of mycoviruses. Further details are given by Day & Dodds (1979).

A major problem with research on fungal viruses is that they are difficult to transmit from one fungus to another. The most effective way, which is doubtless the most common way in nature, is when hyphae fuse with one another by localized breakdown of their walls at the point of contact and cytoplasmic continuity is thereby achieved. This process is termed *anastomosis* and it is commonly seen within a single colony on an agar plate. However, it is under complex genetic control and in many fungi we can recognize *compatibility groups* between which successful anastomoses do not occur. In an incompatible pairing the hyphae of both partners die at the point of fusion and in these conditions it seems likely that the efficiency of virus transfer is much reduced. Reliance on this mode of transmission thus creates problems in studying the effects of viruses, because the effects of other cytoplasmic determinants cannot be excluded; it also creates practical problems in microbial control because of compatibility barriers. In France the use of hypovirulent strains of *Endothia* as control agents is apparently very successful. In the USA, however, there were initial difficulties in using the European strains because they were incompatible with the American ones. About 50 different compatibility groups were identified in the USA (involving an estimated six nuclear gene loci) and it was first necessary to transfer the ds-RNA from hypovirulent European strains into appropriate native strains. Recent research suggests that cocktails of the newly formed hypovirulent strains can be used successfully for control in the field.

It is interesting that most fungal viruses appear to cause latent infections, involving few if any behavioural changes in the host. The virus of *Endothia* is thus unusual, although viruses also cause a degenerative disease of the cultivated mushroom *Agaricus bisporus*, and ds-RNA is associated with a similar degenerative disease of *Rhizoctonia solani*. Other such examples may yet be found and in this respect we can return to the take-all fungus because the first suggestion that any virus might cause hypovirulence was made for this fungus in 1970. French workers reported that isolates from long-term cereal sequences were less virulent than those from short-term sequences and that they also contained isometric virus-like particles. It was therefore suggested that TAD is due to a change to hypovirulence in the pathogen, and it will be recalled that a deficiency in the pathogen would be consistent with our current knowledge of TAD. The suggestion has largely been discounted by subsequent workers for two reasons: first, hypovirulent strains are relatively common in both short and long cereal sequences, as are the virus-like particles; second, detailed work has failed to show any simple relationship between virulence of isolates and their virus content. These may be good enough reasons to question the thoroughness of the French work but they do not invalidate the idea. Taking them in turn, the least virulent isolates probably remain undetected in soil because there is no satisfactory selective medium for isolating *G. graminis* (instead a host infection assay must be used), and the relationship between virulence and virus content is complicated by the fact that at least three sizes of isometric virus-like particle are found in *G. graminis* and there may be other ds-RNA components that are not encapsidated. Added to these points, the discovery of a different type of virus particle in *Endothia* as a result of novel preparation methods suggests that it is too early to discount viruses as a possible cause of TAD.

Suppression of *Rhizoctonia solani* Both natural and induced suppressiveness to *R. solani* have recently been discovered and a central role of *Trichoderma* spp. has been demonstrated in these cases. Naturally suppressive soil from Colombia, South America, contained a high population of *T. hamatum*, and a similar population level added to conducive soil caused it to be suppressive. In the other case (induced suppressiveness) a conducive soil could be made suppressive by repeated cropping of radish or cucumber seedlings in the presence of *R. solani*; this was accompanied by an increase in the population of *T. harzianum* in direct proportion to the degree of suppressiveness. Thus two different *Trichoderma* spp. have been implicated in these cases. The second example is interesting because a modification of Koch's postulates was used to prove the role of *Trichoderma* (Liu & Baker, 1980). With the appropriate reference to Koch's postulates in brackets: (1) *Trichoderma* was constantly associated with suppressiveness (constant association of the suspected causal agent with the disease); (2) *Trichoderma* was isolated at high frequency from hyphal mats of *Rhizoctonia* buried in suppressive soil (isolation of the suspected causal agent); (3) addition of *Trichoderma* to conducive soil at the levels found in suppressive soil caused suppressiveness to develop (reinoculation to reproduce the disease); (4) *Trichoderma* was subsequently reisolated from hyphal mats of *Rhizoctonia* buried in the newly suppressive soil (reisolation and comparison with the introduced organism).

We have seen that *Trichoderma* spp. are microbial control agents in other circumstances; for example, against *Nectria* and *Stereum* on pruning wounds and against *Armillariella* after soil fumigation. In general they are extremely 'aggressive' saprophytes: they have high growth rates, they can utilize a range of polymers like cellulose and hemicelluloses as substrates, and they produce a range of antibiotics active against fungi and bacteria (though not all of these attributes are shared by all species). In addition, some *Trichoderma* spp. are *mycoparasites* – they parasitize other fungi – as evidenced by the fact that they coil around and penetrate the hyphae, they produce wall-lytic enzymes like chitinase and glucanases which can destroy other fungi, and they can use the cell walls of other fungi as sole carbon and energy sources. On this basis it seems reasonable to assume that control of other fungi is brought about at least in part by parasitism. This view is possibly correct, but caution is needed. The alternative suggestion has been made that coiling around the hyphae of other fungi is a means of establishing very close contact so that antibiotics can have the greatest effect. Even this may be unnecessary, because there is one essential feature of an aggressive mycoparasite that is unusual among the fungi as a whole – the ability to grow into colonies of other fungi and thus inevitably to create nutrient-stress. Most fungi do not do this;

rather they are inhibited by the general metabolic by-products of other fungi, loosely termed 'staling factors'. We saw in chapter 6 (pp. 63–4) that autolysis is a common response to nutrient-stress, so it seems likely that the mycoparasites (and there are several of them) could often bring about the destruction of other fungi simply by stressing them. Returning to the example of suppressiveness mentioned above, the fact that *Trichoderma* colonized hyphae of *Rhizoctonia* added to soil is not surprising. Hyphae added to soil in the absence of a suitable substrate are likely to autolyse (see above, pp. 63–4) and if *Trichoderma* (a nutritional opportunist) is present it is likely to colonize the hyphal remains. This is not to dispute the central role played by *Trichoderma* in suppressiveness. It is merely to show that microbial interactions in soil are extremely complex, and the obvious explanation for a phenomenon may not necessarily be the correct one.

Decline of *Heterodera avenae*, the cereal cyst nematode Nematodes or 'eel-worms' are a characteristic component of the soil microfauna. Many of them feed on bacteria and fungal spores and thus are saprotrophic (saprophytic) but several are plant-parasitic and highly specialized for their particular hosts. One such group are the *cyst nematodes*, *Heterodera* and *Globodera* spp, of which *H. avenae* is a specialized parasite of cereals and grasses. The young female larva of *H. avenae* penetrates the host root just behind its tip; the head region becomes embedded inside the root endodermis where the host cells swell into nutrient-rich giant cells in response to invasion, and the rest of the nematode body lies in the root cortex. As it matures into an adult the female swells into a lemon-shape and ruptures the cortex, so that it protrudes from the root. It is then fertilized by wandering males and its uterus becomes filled with eggs which begin to develop into larvae (Figure 18). Then the female dies, development of the larvae is halted and the female cuticle is transformed into a brown leathery *cyst* which can persist for a considerable time in soil. The cycle is completed when a host root passes nearby and stimulates some of the larvae to continue development. These then escape from the cyst and infect the root.

Once established in a site, *H. avenae* is difficult to control by crop rotations because of the long survival of the cysts, but fortunately it is subject to a decline during cereal monoculture. The egg population falls from a peak of about 70/g soil to about 10/g, at which level the nematode does not cause noticeable crop losses. Repeated analyses of decline soils have shown that few adult females survive to form eggs, and this mortality is associated with a high incidence of fungal infection of the females. Several fungal parasites of nematodes are more common in decline than non-decline soils, but *Nematophthora gynophila* is thought to be the most important cause of death of the females, and a second fungus *Verticillium chlamydosporium* seems to be an important parasite of the eggs themselves.

N. gynophila is a zoospore-forming fungus and is placed in the fungal group *Mastigomycotina* (class *Oomycetes*) (Figure 18). Several lines of evidence point to its involvement in decline. For example, decline is always associated with a high incidence of infection of females by it; the decline phenomenon is destroyed by soil treatment with formalin which kills *N. gynophila*; soils in which *H. avenae* is declining can still support increasing populations of other nematodes like the potato cyst nematode *Globodera rostochiensis*, which is not parasitized by *N. gynophila*, and *H. avenae* itself can increase in decline soils in pots in the glasshouse if they are not heavily watered, which is consistent with a role of a

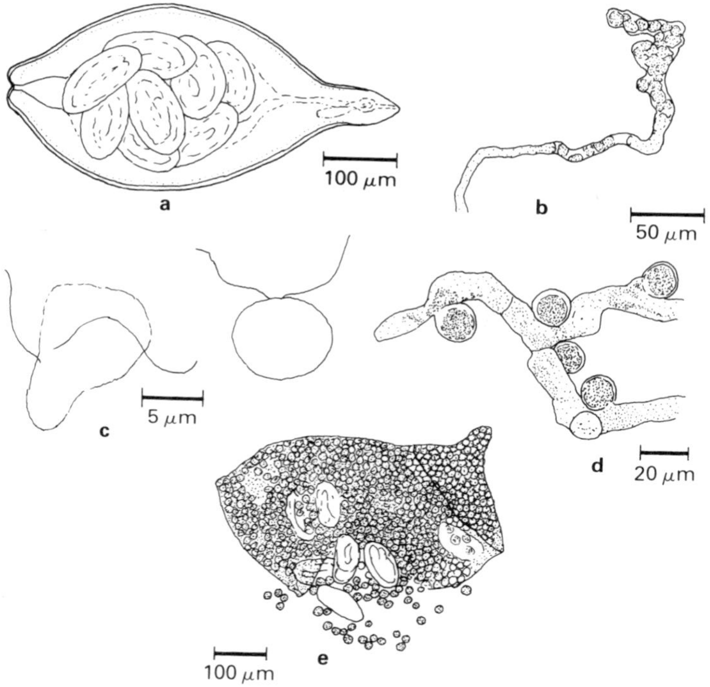

Fig. 18 Cereal cyst nematode, *Heterodera avenae*, and a nematode-parasitic fungus, *Nematophthora gynophila*. **a** Adult female nematode containing embryonated eggs. **b–e** Stages in the life-cycle of *N. gynophila*.
After infection of the host some hyphae develop sporangia at their ends (**b**) and within these the cytoplasm cleaves into motile, flagellate zoospores (**c**). Alternatively the fungus produces thick-walled resting spores (oospores) (**d**) which eventually fill the cyst (**e**). Note that the eggs are not infected. (Drawings **b–e** were traced from photographs in Kerry & Crump (1980) *Transactions of the British Mycological Society* 74: 119–25, by kind permission of the authors and the publisher.)

zoospore-forming fungus in decline because zoospores need free water in order to swim. The literature on this subject can be traced from a recent paper by Kerry, Crump & Mullen (1980), though it should be noted that *N. gynophila* was referred to as an 'Entomophthora-like fungus' in all papers before that date. More generally, Mankau (1980) has reviewed the potential of various fungi, bacteria and viruses for control of nematodes, and this review includes information on the nematode-trapping fungi – a group with some truly remarkable adaptations for the capture of nematodes.

Decline and suppression: summary

No single mechanism can be used to explain all cases of decline and suppression, though there are many similarities between these phenomena, as mentioned at the start of this chapter. Decline is interesting because it must involve an increase in the population or activity of a control agent that is *initially present* in a site. Moreover, in the cases that we have seen the pathogen itself must *selectively favour* the multiplication of the control agent. This is easy to reconcile with cases in which the control agent grows in or on the pathogen; it is more difficult to explain in the case of, for example, take-all decline and the suggested involvement of pseudomonads.

Perhaps only a few cases of decline still remain to be discovered because it is a sufficiently spectacular reversal of a disease trend to be recorded. Naturally suppressive soils, however, may be much more common than is known to date. Even so, there are soils and sites in which no natural suppression or decline has been found and in which disease remains unacceptably high during monoculture. It is tempting to speculate that this is because the resident soil microflora does not contain the necessary control agents. If these can be identified where they occur they might be introduced elsewhere, just as hypovirulent strains of *Endothia* have been introduced in the USA and some insect-pathogenic microbes have been introduced for pest control in various parts of the world (see above, chapters 2 to 4).

The role of mycorrhizas in disease control

Mycorrhizas are symbiotic relationships between plant roots and fungi. The fungus parasitizes the plant in so far as it derives a source of carbon nutrients from it (p. 43) but in return it enhances the uptake of mineral nutrients from the soil, perhaps largely because the fungal hyphae that ramify from the root greatly increase the surface area for absorption. The best known mycorrhizas are the *ectotrophic* or *sheathing* mycorrhizas of forest trees, formed by several members of the *Ascomycotina* and *Basidiomycotina* (the higher fungi) (Figure 19). In this type the main nutrient absorbing branches of the roots are completely surrounded by a sheath of fungal tissue up to 80 μm thick; fungal hyphae radiate from this into the soil, and other hyphae penetrate the root but only *between* the cells, forming a *Hartig net*. This is the type of mycorrhiza to be discussed below. However, the most common type does not have these features and is much less conspicuous. It is termed the *vesicular-arbuscular* (V-A) mycorrhiza: it is formed by members of the *Zygomycotina* (*Glomus* and *Gigaspora* spp.) and it occurs on almost all types of plants in almost all environments, including agricultural ones. There is no sheath around the roots, but fungal hyphae grow into the root cortex where they penetrate the individual cells, forming intricately branched *arbuscules* (the sites of nutrient exchange) and swollen *vesicles* (presumed storage bodies). In addition to these two mycorrhizal types, others are found on orchids and ericaceous plants; indeed there are very few plants either in nature or in agriculture that do not have mycorrhizal associations. The question whether or not mycorrhizal fungi can protect against disease therefore assumes great importance.

The best evidence for a role of mycorrhizal fungi in protecting plants against disease concerns the sheathing mycorrhizas of pine trees, and the pathogen *Phytophthora cinnamomi* (Marx, 1975). This fungus attacks the young feeder roots of a wide range of trees, causing extensive death of these roots and leading to dieback and even death of the shoot system. As shown in Table 7, when roots of pine seedlings are exposed to spores of *Phytophthora* in experimental conditions

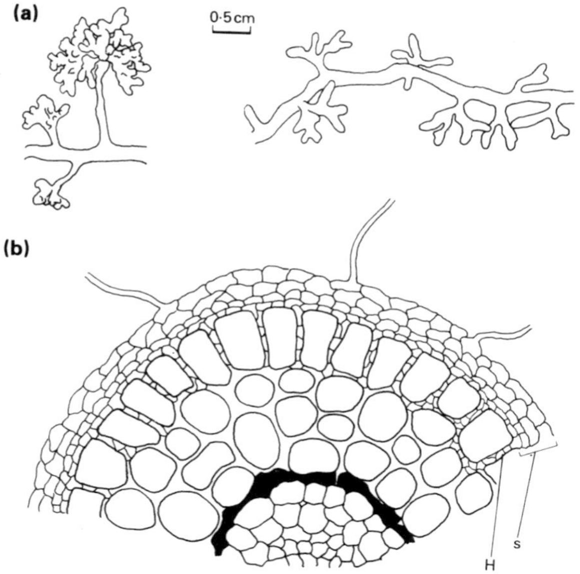

Fig. 19 Sheathing (ectotrophic) mycorrhizas of forest trees. **a** Typical dichotomous branching of short roots of pine when infected by mycorrhizal fungi; sometimes repeated branching leads to a coralloid appearance of the roots. **b** Part of a cross-section of a mycorrhizal root, showing the fungal sheath (s) from which hyphae radiate into the soil, penetration of the fungus between the outer cortical cells to form a Hartig net (H), and dark tannin-filled cells around the vascular cylinder – a common host response to infection.

the non-mycorrhizal roots on the seedlings are severely infected whereas the mycorrhizal roots remain healthy. This remarkable degree of protection is found irrespective of the fungus involved in the symbiosis. However, a further point is raised by the second column of figures in Table 7, which refer to infection of non-mycorrhizal roots in close proximity to the mycorrhizal roots. On

Table 7 Infection of mycorrhizal and non-mycorrhizal roots of pine seedlings by *Phytophthora cinnamomi*. (From Marx & Davey, 1969.)

Mycorrhizal fungus	Per cent infection by *P.cinnamomi*		
	Mycorrhizal roots	Nearby non-mycorrhizal roots	Distant non-mycorrhizal roots
None (control)	–	100	100
Leucopaxillus cerealis	0	25	100
Laccaria laccata	0	100	100
Pisolithus tinctorius	19*	100	100

root systems with *Leucopaxillus cerealis* as the fungal symbiont these nearby roots are significantly protected from infection by *Phytophthora*, but not so when the mycorrhizal fungus is *Pisolithus tinctorius* or *Laccaria laccata*. Protection of the mycorrhizal roots themselves is easily explained: the thickness of the fungal sheath presents a significant barrier to penetration by the pathogen, which is present on the root surface as relatively small spores (a few micrometres in diameter) with minimal nutrient reserves; also the fungus comprising the sheath must use a substantial part of the nutrients that would otherwise leak from the root into the soil and stimulate the growth of the pathogen. Protection of nearby non-mycorrhizal roots cannot be explained in this way, however, and Marx and his colleagues have shown that *Leucopaxillus* produces antibiotics that are active against *Phytophthora*. They are polyacetylenic compounds termed *diatretynes*, and the most potent is *diatretyne nitrile*, shown below, which completely inhibits germination of *Phytophthora* spores at a concentration of 2 ppm.

diatretyne nitrile: $HOOC{-}CH{=}CH{-}C{\equiv}C{-}C{\equiv}C{-}C{\equiv}N$

Pisolithus and *Laccaria* do not form antibiotics active against *Phytophthora*. So it is suggested that *Leucopaxillus* has an extra protective effect on adjacent non-mycorrhizal roots because its antibiotics either are active and diffuse in the root zone or are absorbed by the plant and translocated to other (non-mycorrhizal) roots.

Much field evidence supports the view that sheathing mycorrhizas can significantly protect tree roots against infection by pathogens like *P. cinnamomi*, though not against pathogens of the woody tissues, like *Heterobasidion* and *Armillariella*. This adds an extra dimension to the current practice of artificially establishing selected mycorrhizal fungi on tree seedlings in nurseries. The methods for such establishment are becoming routine, though at present they are used mainly for trees destined for 'difficult' sites like coal bings and other mining wastes, because trees naturally and rapidly develop their own mycorrhizal associations in less hostile environments.

There is much less evidence for a major role of V-A micorrizal fungi in protecting plants against disease, though this has much more agricultural relevance. The subject was recently reviewed by Schönbeck (1979). Although exceptions occur and the subject has received little attention to date, the following generalizations seem to hold. Infection of roots by pathogenic fungi is reduced, but only in the vicinity of mycorrhizal infection and sometimes only in the actual root cells occupied by the mycorrhizal fungus. Infection of roots by nematodes is decreased. Infection of both roots and shoots by viruses is *increased*, perhaps because of the extra available phosphorus in cells of plants with mycorrhizas. Infection of shoots by fungi is *increased*. Thus, overall there seems to be little net benefit of V-A mycorrhizal infection on disease incidence; but it must be said that mycorrhizal plants can be much more vigorous than non-mycorrhizal ones and therefore can perhaps tolerate more disease.

Disease control by induced host resistance

Plants respond to infection or attempted infection in various ways which, together with pre-formed chemical and physical barriers, constitute the plant's resistance mechanisms. The details need not be given here, but we can consider the potentia for microbial control based on these resistance reactions. This potential exists, as evidenced by the fact that some of the more specialized pathogens avoid being recognized as 'foreign' by the plant and therefore can continue their development within its tissues, whereas in an incompatible combination the parasite elicits a host reaction which effectively prevents further development. Thus it may be possible to induce resistance against a compatible pathogen by prior inoculation with an incompatible one. This would be most useful if the resistance were systemic, i.e. if it were to operate on a whole-plant basis. There are a few such examples from experimental work, but as yet it is unclear whether the phenomenon operates in nature and, if so, whether it could be exploited on a commercial basis. Sequeira (1979) reviewed the recognition events between plants and their parasites, including the example to be discussed below; other papers on the subject frequently appear in the journal *Physiological Plant Pathology*.

Pseudomonas solanacearum is an aggressive bacterial pathogen of solanaceous plants (potato, tomato, tobacco); it causes *wilt diseases* by entering the plant's vascular system through wounds and blocking the xylem vessels with its high molecular weight *extracellular polysaccharide* or 'slime'. In experimental systems, avirulent mutants of the bacterium cause rapid localized cell death and fail to multiply when infiltrated into tobacco leaves. Virulent strains, by contrast, cause no such initial reaction; they multiply in the intercellular spaces of the leaves and only after about 36 h do they begin to cause progressive dying lesions in the leaves. The initial host reaction to the avirulent strains is termed a *hypersensitive response* – ironically it is resistance based on extreme susceptibility or sensitivity – and as the cells die they release toxic compounds, like oxidized phenolics which are potent enzyme inhibitors. This hypersensitivity suggests that the plant has recognized the pathogen. Under the electron microscope the avirulent cells are seen to be closely attached to the plant cell walls and they later become enveloped in a fibrillar material from the plant cells. This does not occur in the case of virulent strains. The recognition is now known to be associated with *lectins* in the plant which bind to the lipopolysaccharide (LPS) of the bacterial cell wall. Purified LPS binds to plant cells in the same way as do avirulent bacteria, and the LPS of both virulent and avirulent strains of *P. solanacearum* behave similarly. In fact, the *only difference* between virulent and avirulent strains is that the virulent ones produce slime whereas the avirulent ones do not. Normally, it seems, the slime prevents the LPS from making contact with the lectins in the plant cell walls, so a hypersensitive response is not elicited. The significance for microbial control is that prior inoculation with an avirulent strain prevents subsequent infection by a virulent one. This resistance can become systemic in some conditions so that localized treatment with an avirulent strain or even with LPS of a virulent strain can protect other parts of the plant from infection by virulent *P. solanacearum*. The same *elicitor* (LPS) is involved in both the local and the systemic resistance reaction, but how the systemic effect is brought about is completely unknown.

The response to *Pseudomonas solanacearum* is not an isolated case because, for example, local infection of tobacco leaves by tobacco mosaic virus can bring about a similar systemic resistance to other viruses or fungal pathogens. This *acquired systemic resistance* is different from cross-protection discussed in chapter 5 (pp. 51–4). It is interesting and potentially useful because it is non-specific, operating against a wide range of pathogens. This suggests the involvement of a general host reaction – perhaps a stress response – though the actual resistance mechanism against each pathogen may differ and may simply be enhanced by the underlying stress response. As mentioned earlier, it is unclear whether this could be exploited commercially or whether it operates already in the field. For example, it might be a component of some of the methods of microbial control already discussed (*Agrobacterium* and *Phialophora*). Certainly, plants are frequently exposed to attempted invasion by incompatible parasites, so it is possible that induced resistance, either systemic or localized, is part of the 'background' microbial control in nature.

Summary

Microbial control of plant pathogens is probably widespread in nature and has often been exploited unknowingly in agricultural practice. In some cases soils are naturally suppressive to particular pathogens, and this has been traced to a role of *Trichoderma* spp. in soils suppressive to *Rhizoctonia solani*. In other cases a disease declines after reaching a peak during monoculture; this can be due to mycoviruses (*Endothia* and perhaps take-all decline), other parasites (*Nematophthora* on cereal cyst nematodes) or antagonistic micro-organisms (pseudomonads in the case of take-all decline). Mycorrhizal fungi can protect plants against pathogens; especially the sheathing mycorrhizal fungi of forest trees (e.g. *Leucopaxillus cerealis*) which can protect against feeder root pathogens like *Phytophthora cinnamomi*. Perhaps saprophytes and incompatible plant parasites elicit host resistance reactions in plants in the field and contribute to the background level of disease control. These systems are additional to those discussed in chapter 6 and are currently exploited without the need for special manipulations.

References

Cook, R. J. and Rovira, A. D. (1976). The role of bacteria in the biological control of *Gaeumannomyces graminis* by suppressive soils. *Soil Biology and Biochemistry* 8: 269–73.

Day, P. R. and Dodds, J. A. (1979). Viruses of plant pathogenic fungi. In *Viruses and Plasmids of Fungi* (Ed. P. A. Lemke). Dekker, New York.

Kerry, B. R., Crump, D. H. and Mullen, L. A. (1980). Parasitic fungi, soil moisture and multiplication of the cereal cyst nematode, *Heterodera avenae*. *Nematologica* 26: 57–68.

Liu, S. and Baker, R. R. (1980). Mechanism of biological control in soil suppressive to *Rhizoctonia solani*. *Phytopathology* 70: 404–12.

Mankau, R. (1980). Biological control of nematode pests by natural enemies. *Annual Review of Phytopathology* 18: 415–40.

Marx, D. H. (1975). The role of ectomycorrhizae in the protection of pine from root infection by *Phytophthora cinnamomi*. In *Biology and Control of Soil-Borne Plant Pathogens* (Ed. G. W. Bruehl). American Phytopathological Society, St. Paul, Minnesota.

Marx, D. H. and Davey, C. B. (1969). The influence of ectotrophic mycorrhizal fungi on the resistance of pine roots to pathogenic infections. III. Resistance of aseptically formed mycorrhizae to infection by *Phytophthora cinnamomi*. *Phytopathology* 59: 549–58.

Schönbeck, F. (1979). Endomycorrhiza in relation to plant diseases. In *Soil-Borne Plant Pathogens* (Ed. B. Schippers and W. Gams). Academic Press, London.

Sequeira, L. (1979). Recognition between plant hosts and parasites. In *Host-Parasite Interfaces* (Ed. B. B. Nickol). Academic Press, New York.

8 Conclusion

Much of this book has been concerned with specific examples of pest and disease control; more than twenty examples have been considered in detail and the discussion has ranged more widely to cover some general mechanisms of control that have not yet found commercial application. But twenty examples is not many, and even if it had been 200 it would still not nearly approach the scale on which other methods like chemical control and plant breeding are used in practice. So are we to conclude that microbial control is less important in general than other methods? This is the basic question to be considered in this chapter and it provides an opportunity to review the topics in previous ones.

First, it must be said that the 'balance' between pest and disease control in the book may have been misleading. Plant diseases are caused primarily by micro-organisms which live in communities with other micro-organisms and are affected by them. So for plant pathogens most examples of *biological* control are also examples of *microbial* control, and we have seen that this can be brought about in various ways. Most plant pests, however, are insects or other small animals and they live in communities with other insects. The only direct way in which they are affected by micro-organisms is when the micro-organisms cause diseases; all the examples of microbial control of pests have involved insect-pathogenic microbes. In fact, the *biological* control of pests is brought about mainly by other small animals which act as predators or parasites; there are numerous successful examples of this in commercial practice but they are outside the scope of this book. To be realistic, therefore, we should rephrase the question above: is *biological control* in general less important than other control methods?

The answer would have to be 'Yes' except for one overriding factor. Biological control probably operates much of the time in most plant communities but goes unnoticed. The evidence for this is not hard to find even if we restrict the discussion to microbial control agents. Thus, insect diseases caused by bacteria, viruses, fungi and protozoa occur naturally without man's interference; we saw for example in chapter 4 how *Entomophthora* spp. frequently cause natural epizootics in aphid populations in the field. Similarly, *Peniophora gigantea* naturally competes with *Heterobasidion* for pine stumps in British forests (chapter 5); disease-suppression and decline occur naturally in agricultural practice (chapter 7); mycorrhizal fungi are the rule rather than the exception in natural and agricultural soils (chapter 7), and non-specific microbial control mechanisms like germination-lysis (chapter 6) probably occur in all soils. In fact, most of the successful applications of microbial control in commercial practice have stemmed from the recognition of these natural phenomena. Classic examples are the introductions of insect pathogens into countries where they did not occur. In chapter 3 we saw how the introduction of baculoviruses into North America controlled serious problems caused by sawflies in the forests; it could be argued justifiably (but without proof) that sawflies are not a problem elsewhere – especially in areas where they are indigenous – because of the activities of natural enemies. The same appears to be true for the Japanese beetle, which was brought under control in the USA by the use of milky disease bacteria (chapter 2).

If this is true in general, then we can argue that biological control is one of the most important factors limiting damage by pests and pathogens and that we see such damage only when for some reason biological control is not operating effectively. This is not special pleading. The same type of argument has led plant breeders back to the centres of origin of our major crop plants, to find disease-resistance genes in the wild ancestors of our crops. In practical terms it has major implications for the future of microbial control: if we can identify the organisms that naturally control pests and pathogens in crops or geographical areas where damage is seldom seen, then we can try to introduce the agents into other areas. This has already been done successfully for insect pests, but surprisingly it has seldom been tried for plant pathogens; an obvious exception concerns virus-infected hypovirulent strains of *Endothia parasitica* (chapter 7) which have recently been introduced into the USA from Europe.

One further point should be made in relation to 'natural' microbial control. Relatively little attention has been given to the possible involvement of micro-organisms in the success of chemical control measures or even plant breeding. We saw in chapter 4 that insecticides can stress insects and make them more susceptible to microbial pathogens; in chapter 6 we saw a similar example in the use of carbon disulphide against the pathogen *Armillariella mellea*, and a possible secondary role of competing micro-organisms following use of the fungicide benomyl against *Pithomyces chartarum*. Perhaps one day the chemical companies will boast that their products 'help to restore the natural balance' rather like the current range of hair shampoos.

Special constraints on microbial control

Turning now to the practical application of microbial control, there are several special constraints that limit its commercial use. They fall basically into three categories.

1 Ideally a control measure should involve a minimal departure from current agricultural practices. The reason is that pest and disease problems are only two of the many factors that a grower needs to consider; often they are not the critical limiting factors in crop production, and even if they are so they cannot always be predicted. Disease resistant cultivars are attractive because they are cheap and involve no special manipulations. Chemical control is similarly attractive because it can be used as and when required and usually has immediate effects. Unfortunately, microbial control agents seldom have these advantages. Good examples, however, include *Bacillus thuringiensis* which is used as a microbial insecticide (chapter 2), *Agrobacterium radiobacter* which is applied as a pre-planting spray or dip to control crown gall (chapter 5) and attenuated strains of viruses like TMV and tristeza virus which protect plants against virulent strains (chapter 5). Apart from such cases the use of microbial control agents is based on special circumstances. For example, they can give longer term control than most chemicals, which compensates for the extra inconvenience. Introductions of *Bacillus popilliae* and baculoviruses (chapters 2 and 3) are amongst such cases, as is the use of *Beauveria bassiana* against Colorado beetle (chapter 4) and of *Peniophora* against *Heterobasidion* (chapter 5). In the last case we saw how the treatment is still inconvenient and attempts are being made to introduce spores of *Peniophora* into chainsaw oil to simplify the method. In yet other cases microbial control is used because other preferred methods are less effective. Thus, *Verticillium lecanii* is likely to be used increasingly for aphid control in glasshouses because several aphids have developed resistance to the commonly used aphicides (chapter 4); *Bacillus thuringiensis* is used on lettuce and other crops in Arizona and California because the most effective chemicals cannot be applied within one month of harvest (chapter 2); baculoviruses are used against the rhinoceros beetle in the South Pacific because the virus is spread by the beetle itself and chemicals cannot be introduced effectively into the breeding sites (chapter 3); many root diseases and root-infecting pathogens are controlled by microbial agents because fungicides cannot easily be introduced into the root zone of growing crops (chapters 5 and 6).

It is notable that some of the best examples of microbial control relate to glasshouse-grown crops or other crops which have a high market value and are labour-intensive (i.e. there is traditionally a large degree of handling of the plants). The use of aerated steam and soil fumigants falls into this category (chapter 6), as does the protection of stumps and pruning wounds (chapter 5). In glasshouses in particular there is a strong tradition of innovative cropping practices, made possible or even necessary by the high costs involved with these crops and the large potential profits.

2 Microbial control is more difficult to achieve than are some other control methods because it involves the use of living organisms. There are problems of inoculum production, especially if the control agent proliferates (baculoviruses) or sporulates (*Bacillus popilliae*) only *in vivo*; *Nematophthora gynophila*, the cause of decline of cereal cyst nematode, would also fall into this category if it were to be applied as a control agent (chapter 7). There are problems of inoculum storage, especially with fungal spores (chapter 4); but these problems have been overcome by refrigeration (e.g. agar plates and liquid inocula of *Agrobacterium radiobacter* and spores of insect-pathogenic fungi), by freeze-drying (*Agrobacterium*) or by using osmotically active solutions (*Peniophora*). There are problems of application and persistence of inoculum, especially in field conditions where the environment cannot be controlled; the viability of many micro-organisms is lost rapidly on exposure to ultraviolet radiation, drought or high temperatures. And there is the general problem of predictability which becomes increasingly more difficult with increasing complexity of the interaction between the control agent, the target species and the rest of the microbial community. This problem can of course be minimized by adequate testing before a process is made available for use. However, if an unforeseen problem should arise the grower himself cannot be expected to deal with it, so greater access to specialist personnel is needed than with most other control measures.

3 The last special constraint concerns the development stages of a microbial control measure. Living organisms and natural processes cannot be patented, although this problem can be circumvented by patenting a means of using microbes for a specific purpose and in the case of genetically modified strains. It is correspondingly difficult for a commercial organisation to recoup its development costs. Furthermore, the limited potential market for many microbial control agents and the persistence of the control in some cases are disincentives to large commercial concerns. Nevertheless, several small companies are involved in inoculum production for microbial control; examples are Fairfax Biological

Laboratory Inc. who produce spore powder of *Bacillus popilliae* in the USA, and Ecological Laboratories Ltd of Dover, who market sachets of *Peniophora* spores in Britain. Also some of the larger fermentation industries are showing interest in producing selected microbial control agents on a pilot scale for field testing; spores of *Trichoderma* spp. are obvious candidates in this respect because of the involvement of these fungi in several microbial control processes (chapters 5, 6 and 7).

In practice, microbial control measures are conceived in universities and Government research establishments, and all stages in their development – from the fundamental studies to the final field testing – are carried out by people in these institutions.

Epilogue

It is pointless to try to predict the future of microbial control, simply because each control measure must be 'tailor made' and it depends on the rate of increase in our knowledge and understanding of the environment in which crops grow. Nevertheless, microbial agents lend themselves so readily to use in integrated control programmes that this seems undoubtedly the main area for future development.

The notable successes to date can serve as models for future work, and there can surely be no doubt that microbial control, and biological control in general, has been placed firmly on the map in the last twenty years or so. Perhaps the most satisfying and substantial achievements on the road to microbial control have been the insights into microbial behaviour in nature, the discovery of entirely new types of microbial interaction, and the marriage of 'pure' and 'applied' biology which is essential to the rational development of microbial control. It is this inextricable link between pure and applied aspects of biology that I hope this book has managed to demonstrate.

Glossary

Several terms are discussed in detail at the first mention in the book, referred to by the page numbers in parentheses. The following should not be regarded as strict definitions, merely as explanations.

Actinomycete: filamentous bacterium.
Antagonism: direct detrimental effect of one organism on another (43).
Ascomycotina: one of the main groups of fungi; the hyphae have septa (cross walls), the asexual spores are conidia, and sexual spores are formed within an ascus (sac-like cell).
Attenuated: with reduced virulence.
Avirulent: non-pathogenic (43).
Basidiomycotina: one of the main groups of fungi; the hyphae have septa (cross walls), asexual reproduction (where present) is by conidia, and sexual reproduction often results in development of a fruiting body (e.g. mushroom or toadstool).
Basidiospore: sexual spore of the Basidiomycotina.
Cadaver: corpse.
Canker: a sunken disease lesion in woody tissues; the tissues disintegrate outside the xylem cylinder, and no new xylem is produced beneath the lesion.
Chlamydospore: fungal resting spore formed by deposition of a thick secondary wall within an existing vegetative cell or occasionally within a conidium.
Competition: indirect detrimental effect of one organism on another (43).
Conidium: asexual spore of the Deuteromycotina, Ascomycotina and Basidiomycotina; it is *not* formed within a sporangium.
Constitutive dormancy: failure of a spore to germinate in suitable environmental conditions because of endogenous factors.
Cuticle: outer covering of insect, composed of a thin hard *epicuticle* and an underlying softer *procuticle*.
Damping off: disease of seedlings in which the tissues develop a water-soaked appearance and die.
Depsipeptide: cyclical peptide.
Deuteromycotina: one of the main groups of fungi; the hyphae have septa (cross walls), asexual reproduction is by means of conidia, and sexual reproduction is absent or rare (see p. 32).
Disomic: relating to a chromosome pair.
Endodermis: a specialized layer of cells between the cortex and the vascular (conducting) system of plant roots or stems.
Epizootic: equivalent to an epidemic (6).
Globose: rounded, globe-like.
Haemolymph: blood.
Hepatotoxin: toxin with a primary effect on the liver.
Hypha: the basic unit of a fungus; a fungal filament.
LD_{50} (lethal $dose_{50}$): the dose needed to kill 50% of individuals.

Lectin: material involved in recognition reactions; by definition, a proteinaceous material that agglutinates red blood cells.

Lysis: enzymic breakdown (43).

Maintenance energy: energy needed to maintain viability of a cell.

Mastigomycotina: one of the main groups of fungi, characterized by the possession of motile flagellate spores (zoospores) formed by cytoplasmic cleavage within a sporangium.

Mycelium: collective term for a network of hyphae.

Mycoparasite: fungus parasitic on another fungus.

Parasite: organism that obtains nutrients from a living host (43).

Pathogen: disease-producing organism (usually a parasite) (43).

Phloem: part of a plant's vascular system, in which sugars are transported.

Plasmid: circular piece of bacterial DNA independent of the chromosome (48–51).

Predator: organism that captures food (prey) (43).

Septicaemia: presence of pathogenic bacteria in the blood.

Sympodial: a form of growth in which the growing point is used to produce a spore etc., and a new growing point is formed beneath it.

Systemic: generalized; a fungicide that can be applied to a plant surface and is taken up and redistributed via the xylem.

Tumorigenic: producing tumours.

Vascular system: (of a plant) the xylem and phloem and associated supporting tissues.

Virulence: ability of a pathogen to cause disease.

Xylem: water-conducting tissues of a plant.

Zoospore: Motile flagellate fungal spore.

Zygomycotina: one of the main groups of fungi; the hyphae characteristically lack cross walls, asexual spores are non-motile but are formed in a sporangium like those of Mastigomycotina, and sexual reproduction results in a spiny thick-walled resting spore (zygospore).

Index

Index